SAVING THE CHESAPEAKE

SAVING THE CHESAPEAKE

THE HISTORY OF A MOVEMENT

ANDREW S. RAMEY

RIVANNA BOOKS
UNIVERSITY OF VIRGINIA PRESS
Charlottesville and London

The University of Virginia Press is situated on the traditional lands of the Monacan Nation, and the Commonwealth of Virginia was and is home to many other Indigenous people. We pay our respect to all of them, past and present. We also honor the enslaved African and African American people who built the University of Virginia, and we recognize their descendants. We commit to fostering voices from these communities through our publications and to deepening our collective understanding of their histories and contributions.

Rivanna Books
An imprint of the University of Virginia Press
© 2025 by the Rector and Visitors of the University of Virginia
All rights reserved
Printed in the United States of America on acid-free paper

First published 2025

1 3 5 7 9 8 6 4 2

Library of Congress Cataloging-in-Publication Data

Names: Ramey, Andrew S., author.
Title: Saving the Chesapeake : the history of a movement / Andrew S. Ramey.
Description: Charlottesville : Rivanna Books, University of Virginia Press, 2025. | Includes bibliographical references and index.
Identifiers: LCCN 2024041114 (print) | LCCN 2024041115 (ebook) | ISBN 9780813952659 (hardcover) | ISBN 9780813952666 (paperback) | ISBN 9780813952673 (ebook)
Subjects: LCSH: Environmental protection—Chesapeake Bay (Md. and Va.)—History. | Environmental protection—Chesapeake Bay Region (Md. and Va.)—History. | Water quality management—Chesapeake Bay (Md. and Va.)—History. | Water quality management—Chesapeake Bay Region (Md. and Va.)—History. | Chesapeake Bay (Md. and Va.)—Environmental conditions. | Chesapeake Bay Region (Md. and Va.)—Environmental conditions. | Estuarine pollution—Chesapeake Bay (Md. and Va.)
Classification: LCC TD225.C43 R36 2025 (print) | LCC TD225.C43 (ebook) | DDC 363.73/70916347—dc23/eng/20241119
LC record available at https://lccn.loc.gov/2024041114
LC ebook record available at https://lccn.loc.gov/2024041115

Cover art: Osprey, ajushn/shutterstock.com; sky, lovelyday12/shutterstock.com; background, flownaksala/istock.com
Cover design: Susan Zucker

CONTENTS

List of Illustrations | vii
Preface | ix
Acknowledgments | xiii

Introduction: The American Estuary   1

1. "Save the Bay"   17
2. The Storm   47
3. The Chesapeake Bay Agreement   74
4. Progress and Backlash   104
5. Making a National Treasure   141

Conclusion: Looking Back, Looking Ahead   187

Notes | 197
Bibliography | 221
Index | 231

ILLUSTRATIONS

FIGURES

1. Original "Save the Bay" bumper sticker, 1969 — 19
2. Charles Mathias receiving an award, 1983 — 49
3. Chesapeake Bay Foundation membership, 1970–1983 — 84
4. William Baker testifying before Senate subcommittee, 1983 — 89
5. Signing the Chesapeake Bay Agreement, 1983 — 96
6. Maryland striped bass juvenile index, 1966–1996 — 107
7. Maryland oyster landings, 1982–2004 — 116
8. "Fact Sheet: Detergents, Phosphorous, and the Bay," 1985 — 126
9. Menhaden with lesions associated with *Pfiesteria piscicida* — 153
10. Chesapeake Bay Foundation membership, 1984–2004 — 162

MAPS

1. Chesapeake Bay watershed — 5
2. Key locations in the Chesapeake Bay — 34
3. Rainfall from Tropical Storm Agnes, 1972 — 57

PREFACE

This book is the first comprehensive history of the movement to restore the Chesapeake Bay from its origins in the 1960s through the first decade of the twenty-first century. I sincerely hope that it is not the last history of the Chesapeake restoration movement, because the subject is too important to be limited to a single book. I hope that this book will raise as many compelling questions as it answers and inspire others to pick up the trail where I have left it for either want of evidence, lack of publishing space, or narrative cohesion. Every historian must make difficult choices about what sources to use, whom to emphasize, what the major events of the time were, and then submit the final product for judgment by scholarly peers and the public. These general remarks are true of any good history, but a few specific points are relevant to this chronicle of the Chesapeake Bay's environmental history.

The Chesapeake is a tremendous research subject because the $30 billion joint state and federal restoration initiative encompasses parts of six states that make up the Bay's watershed and includes components as large scale as U.S. Supreme Court decisions and as small scale as backyard stream restoration. In between, multitudes of state, local, federal, and private projects lurk, waiting to be studied. The breadth of issues the restoration effort addresses, ranging from fisheries and wilderness to industrial agriculture and suburban sprawl, is similarly inviting for those curious about environmentalism, public policy, economics, science, and much more. If there is an environmental issue you are interested in, there is likely a Chesapeake case study that can shed some light on it (and you should go research it, right after finishing this book). The challenge, of course, is to knit all these elements together in a coherent narrative and convincing argument. In order to do so, I have been forced to make many difficult decisions about whom and what to include in this volume and whom and what to leave out. The most consequential decisions have been to focus on the role of the federal government and the role of the region's largest environmental group, the Chesapeake Bay Foundation (CBF). Readers already somewhat familiar with the Bay restoration effort may know that state and local governmental agencies, scientists, commercial fishers (watermen, to locals), and other environmental groups have had a profound impact on the course of the Bay's restoration.

I touch on these other actors as often as possible in the text, but because of limitations on space and sources and a desire to provide a rich chronological overview, I have chosen to focus on CBF and the federal government. The latter is undoubtedly significant, and federal records are abundantly available. Additionally, one of the main arguments of this book is that understanding the changing role of the federal government, from a tangential supporter to a central player, is essential to understanding the history and recent successes of the Bay restoration movement. My choice to focus much of the narrative on environmentalists at CBF, however, deserves a slightly more detailed explanation. One of the anonymous reviewers who read an earlier draft of this work rightly pointed out that an unfamiliar reader might conclude that CBF was the dominant player in the Bay's restoration, when in fact it was only one of many stakeholders, including scientists employed by state and federal agencies, as well as universities, other citizens' groups, including the Alliance for the Chesapeake Bay, valuable nonprofit watchdog journalists like those at the *Bay Journal*, and many others. I should also emphasize that these various stakeholders by no means marched in lockstep and often had heated disagreements about the direction of the restoration movement. However, in an effort to explain why the movement as a whole achieved meaningful results for the Bay, I have chosen to focus more on the struggle between supporters and opponents of Bay restoration than on the divisions within the movement.

While I acknowledge CBF as one player among many, it deserves a prominent place in any analysis of the Chesapeake Bay's restoration since it is the regional movement's first and largest environmental organization. As the first chapter of this book details, CBF predated such nationally significant events as Earth Day and the formation of the Environmental Protection Agency and such regionally specific events as the creation of the Chesapeake Bay Program. Moreover, many of the records from CBF's earliest years survive in the Special Collections at the University of Maryland's Hornbake Library. I discovered these records as a graduate student, and this book represents their first scholarly use. Thanks to this remarkable source base, plus other archival records at the University of Maryland and Johns Hopkins University, my own oral histories, and the willingness of William Baker, at the time the president of CBF, to open the organization's internal records to me, I was able to provide an "insider's account" of one of the most important players in the Chesapeake's restoration that provides a throughline from the movement's earliest years in the 1960s into the twenty-first century. This level of detail about other players simply was not possible for reasons of both space in the book and inaccessibility of sources. Two anecdotes will illustrate the challenges facing a historian of the Bay's restoration.

The same anonymous peer reviewer who encouraged me to be clearer about the limits to focusing on CBF acknowledged the struggle to document the history of this movement. The reviewer wrote, "I saw twice in my career large amounts of records simply discarded, ostensibly because there was simply no place to store them in a pre-computer society." This was not an isolated phenomenon; I encountered many examples of lost or destroyed materials in my own research. Often, important documents survived by sheer chance. For example, during an interview with Ann Swanson, then the executive director of the Chesapeake Bay Commission, she mentioned that she had some old files from her time at CBF and asked if I was interested in them. Seeing my obvious interest, she pulled out a wicker basket full of loose-leaf materials about grassroots campaigns she had organized at CBF in the mid-1980s. My jaw nearly hit the floor. These materials, which form a substantial source base for my case study of the phosphate bans in chapter 4 and my understanding of how CBF's grassroots arm worked, were a goldmine for me. I had no idea of their existence, and I stumbled upon them through sheer happenstance. I am sure many more records are out there to be discovered, and I hope the publication of this book will encourage more research into the subject and encourage participants in the Chesapeake's restoration to donate their materials to archives capable of preserving them and making them available to the public.

Thus, the history you are about to read is a product of no small amount of good fortune. It is also the product of much hard work, digging, sleuthing, and help from many people whom I thank more fully in my acknowledgments. While this is the first comprehensive history of the Bay's restoration, it is by no means *the* history of the movement. It is an opening historical salvo, the tip of the iceberg, or to use a more Chesapeake-themed analogy, it is one tributary flowing into the larger literary estuary where many different accounts blend, like saltwater and freshwater, to produce a perspective on the Bay that is as unique and distinctive as the estuary itself. For those interested in reading more about the Bay, as well as environmentalism and public policy, I recommend a careful perusal of the bibliography. To anyone interested in making their own contribution to the subject, I say do it! History is one of the most democratic disciplines; you do not need access to a supercollider or an expensive laboratory to research and write history. Public records belong to the people, and we should all use them to understand our place in the world. I hope this book accomplishes many things, not the least of which is to inform, inspire, and invite readers to learn more about the Chesapeake and all those who have fought to save the Bay.

Andrew S. Ramey
Pittsburgh, Pennsylvania
July 8, 2024

ACKNOWLEDGMENTS

Everything has a history, including this book. It began many years ago as an undergraduate research project; my fascination with the subject and the support of incredible teachers and mentors led me to continue asking questions about the Chesapeake in graduate school, where the bulk of the research for this book took place, forming the core of my dissertation. In the years since, I have returned to my sources, revisited my thinking, revised my prose, and produced this volume, which would not have been possible without a small army of supporters. I will try to name as many of them as I can, with any omissions being unintentional and sincerely regretted.

I am deeply indebted to Bob Doherty, who became my mentor when I was an undergraduate at the University of Pittsburgh. Bob agreed to supervise my summer research project and taught me by his example how to be a good historian and a better person. I was fortunate to have many other great advisers and role models, who encouraged me to pursue graduate school, among them Tony Novosel, who showed me how much fun history can be, and Mike Giazzoni, who constantly pushed me out of my intellectual comfort zone. I am also grateful to Ed McCord, Alec Stewart, and Bernie Hagerty for their guidance, as well as the many lifelong friends from Pitt's Honors College, who helped me learn how to think clearly, critically, and compassionately.

As a graduate student in Carnegie Mellon's superb Department of History I was fortunate to have not one but two dissertation advisers, Joel Tarr and John Soluri. I am always mindful of Joel's sage advice that you cannot jump up a flight of stairs, you have to take them one at a time. John was especially crucial to the development of this book, first as a faculty adviser and then as a friend and colleague. I would not have finished the book without John's encouragement, as well as the occasional coffee-fueled writing sessions. I thank Nico Slate for serving as my third committee member and for helping to shape the subsequent book with his trademark warmth and generosity. Caroline Acker, Paul Eiss, Wendy Goldman, Ricky Law, Kate Lynch, Steve Schlossman, and Joe Trotter have all influenced my approach to thinking, talking, and writing about history. Many tremendous historians overlapped with me while I was a graduate student, Alissa Belotti, Laura Bradshaw, Kevin Brown, Susan Grunewald, Jiacheng Liu, Matt McGrath, Cian McMahon, Cassie Miller, Matt Nielsen, Avigail Oren, Jonathan Stepp, Amund Tallaksen, and Pat Zimmerman among them. I am especially grateful for Kaaz Naqvi's ability to encourage and challenge me at the same time; I could not have asked for more from a colleague and a friend.

During my research for this project, many people took the time to talk with me, on and off the record, about the history, science, and politics of the Chesapeake Bay restoration effort. My thanks to Will Baker, Don Baugh, Bill Goldsborough, C. A. Porter Hopkins, Tom Horton, Minor Lee Marston, Jon Mueller, Patrick Noonan, Russell Scott, Truman Semans, Suzanne Sherwood, Ann Swanson, John Page Williams, Mary Tod Winchester, and Laura Wood. Additional thanks to Will Baker for allowing me access to some of CBF's internal records and to Darlene Warnken for finding all the old files I needed; Tom Horton for generously gifting me a copy of his "Hargis, Cronin, Pritchard" video; Ann Swanson for granting me access to her personal collection of documents from her time at CBF and the Chesapeake Bay Commission. I am deeply indebted to the archivists at the University of Maryland's Special Collections and Johns Hopkins University's Ferdinand Hamburger University Archives for helping me access their collections. I express my gratitude to the University of Toronto Press to reuse in chapter 1 portions of my earlier work, "The Calvert Cliffs Campaign, 1967–1971: Protecting the Public's Right to Knowledge," which appeared in *Nuclear Portraits: Communities, the Environment, and Public Policy*, edited by Laurel Sefton Macdowell in 2017. I would also like to thank Michael Lewis and Creston Long for their leadership in the field of Chesapeake studies; I am especially grateful for the opportunity to speak to the environmental studies community at Salisbury University. Talking to people who care about the Bay helped me sharpen the final argument and message of my book and inspired me to get over the final hurdles to finishing the manuscript. I hope I can return to the Eastern Shore again very soon. I would like to thank my anonymous peer reviewers for their helpful suggestions. I am grateful to the talented people at the University of Virginia Press, including Clayton Butler, Andy Edwards, Angie Hogan, as well as Joanne Allen, who did a fantastic job copyediting the manuscript, catching my mistakes and reining in some of the excesses of my prose. All errors are mine and mine alone. I would like to give special thanks to Mark Mones for championing this project, sharing my enthusiasm for ospreys, and accepting an out-of-the-blue invitation to a Zoom talk about the Chesapeake Bay at UVA's School of Architecture.

Writing a scholarly book is difficult enough, but it is all the harder when you are doing it on the side, giving up weekends and evenings with friends and family because there is a burning story that needs to be told. This book very nearly died a quiet death in 2019, when I was running on fumes and ready to quit. Fortunately, my father-in-law, George Blashford, invited me to be the keynote speaker for an Earth Day event at his church, Market Square Presbyterian, in downtown Harrisburg, Pennsylvania. Speaking to a general audience about the importance of saving the Bay on a holiday born from the environmental movement reenergized me and helped me find my message. Essential to the making of this book have been the love and support of family, from my mom and dad, Marty and Andy Ramey, to my brother, Nick, and the incredible Blashford family I married into; special thanks to Turner Blashford for the maps, which I could not have done without him. Thanks to Sam, Roc, and Rory for the distractions—how did you always know when I needed a break? Most of all, I could not have done this without the love and support of my wife and best friend, Kate Ramey. Thank you for always being there, for helping to bring out the best in me, and helping me to become more of myself than I ever could on my own. Finally, I must acknowledge the Chesapeake Bay and all its defenders, advocates, and protectors for giving me such a worthy research subject. Researching and writing this history has been an incredible honor and a tremendous challenge. I am forever grateful to everyone who has helped me complete it.

SAVING THE CHESAPEAKE

INTRODUCTION
THE AMERICAN ESTUARY

The Chesapeake Bay is the subject of one of the longest-running and most expensive ecosystem restoration projects in the world. Since 1983, state, federal, local, and private sources have combined to spend more than $30 billion to restore the Bay, with half that total coming just since 2014. Other major federal programs, such as the Comprehensive Everglades Restoration Program, with spending in the neighborhood of $10 billion over the past twenty years, and the Great Lakes Restoration Initiative, with a comparatively paltry $4 billion over the past ten years, are dwarfed by the Chesapeake Bay restoration effort in terms of scale, cost, timeline, and the complexity of the politics, ecologies, and economics. Despite the significance of this project for the region and the nation, there has been no comprehensive historical analysis of the origins, evolution, and achievements of the social and political movement to save the Chesapeake Bay.[1]

The sheer amount of money invested would in itself make the Chesapeake Bay a supremely important site for examining American environmental policy at both the state and the federal level. Beyond that, the Bay's cultural, historical, economic, and of course ecological significance is unparalleled. The Chesapeake Bay has been a cradle of American society since 1607, when settlers at Jamestown established the first permanent English settlement in what would become the United States. The estuary was a crucible for the reforging of the nation, first during the War of 1812 and then, much more dramatically, during the Civil War, when the blood spilled by Union and Confederate troops from Gettysburg to Fredericksburg ultimately drained to the Bay. The Chesapeake helped give rise to American economic power in the industrial metropolis of Baltimore and to American sea power in the strategic harbor of Norfolk. As it is the largest estuary in the United States, it is fitting that the nation's capital straddles the Bay's tidal reaches along the Potomac River. Without question, the rich history flowing throughout the Chesapeake Bay region marks it as the American estuary. The efforts to restore the American estuary are only the latest example of grand dramas to play out along its shores.

From an environmental perspective, the Chesapeake Bay is a majestic ecosystem on par with celebrated landscapes in the American environmental pantheon such as Yellowstone, the Adirondacks, or the aforementioned Everglades. Yet the Chesapeake has not occupied a prominent place in the nation's environmental history commensurate with its standing as one of the country's great natural treasures. In part this omission comes from the estuary's lack of the charismatic megafauna or stunning vistas of more iconic national parks. One of the Bay's early environmental advocates, Arthur Sherwood, complained that "the quiet and subtle beauty, the fragile character of Bay country was over-looked by the kind of early amateur naturalist who flocked to the cause of protecting mountain, stream, and giant redwood." To Sherwood and many others, the Bay was a unique blend of natural and human influences, and the long history of Euro-American settlement and subsequent development of industrial agriculture, cities, and sprawl continues to blind observers to the Bay's significance, especially when compared with "pristine" wildernesses, particularly those located in the far western or far northern United States. A southeastern estuary that is home to major cities, military bases, intensive agriculture, and a nearly unbroken swath of suburban development does not fit popular notions of what an iconic American ecosystem should be.[2]

It is one of the chief goals of this book to challenge the bias toward supposedly untouched western landscapes that still persists in the popular and scholarly imagination. While the Bay possesses secluded haunts like Blackwater National Wildlife Refuge, which can evoke all the mystery and wonder of wild nature, and while the Bay is home to animals as charismatic as the most quintessentially American symbol, the bald eagle, to rely upon these examples to justify the Bay's natural significance would be to miss the point and reinforce the blinders this book seeks to remove. The Chesapeake's wild spaces and wild animals are only a small part of the Bay's appeal; what makes it such an important subject to study is that after centuries of supporting the growth of industrial society along its shores, the Bay's ecosystem—wild, tamed, and all degrees in between—endures. It is an "organic machine" that continues to hum with a vibrant energy that flows through the people, plants, and animals in its watershed. The Chesapeake should be celebrated as a great American ecosystem not because some parts of it remain pristinely protected from human interference—such places are as mythical in the Chesapeake as they are anywhere else on earth—but because it offers a rare example of sustainability.[3]

Although incomplete and ongoing, the movement to protect and restore the Chesapeake Bay is a compelling case study because despite coming up short in their goal to fully restore the Bay, environmentalists have managed to sustain the fundamental integrity of the ecosystem in the face of decades of population growth, the expansion of suburbs and exurbs throughout the watershed, ongoing commercial fishing, intensification of agriculture, and a host of other stressors to the Bay's systems. A deep historical analysis of Chesapeake environmentalism—which emerged in the 1960s, evolved by the 1980s to support a joint state and federal restoration initiative known as the Chesapeake Bay Program, and fully matured in the early years of the twenty-first century as a force powerful enough to redirect national policy and significant federal funding toward the Bay—reveals important insights about U.S. politics and the U.S. environmental movement, while chronicling the resuscitation of a major ecosystem. In short, the argument presented in these pages is that the Chesapeake Bay has always been the American estuary, but it took nearly fifty years of activism, collaboration, and a little bit of luck for Chesapeake environmentalists to force the federal government to start treating it that way. While saving the Chesapeake is a significant achievement, one must not overstate the case: the Bay remains impaired and imperiled, by missed opportunities and looming threats. The present and future Bay faces grave challenges, from population growth to climate change, and it is certainly less healthy and less resilient owing to paths not taken.

Nevertheless, a full accounting of the origins and evolution of the Chesapeake Bay restoration effort reveals more wins than losses. Assessing this record is all the more necessary in order to arm present generations with the hard-won knowledge of their predecessors' past efforts.

MAJOR THEMES

Untangling the history of the Chesapeake's environmental movement is daunting, if for no other reason than that so many important institutions contain the name of the estuary in their title. A list of major contributors to the movement includes the Chesapeake Bay Foundation, the Chesapeake Research Consortium, the Chesapeake Bay Commission, the Chesapeake Bay Program, the Chesapeake Bay Trust, and the Alliance for the Chesapeake Bay, among many others that have come and gone. The region's history is also dizzyingly complex because it is so decentralized. Important policies can come from any of three state capitals—Annapolis, Harrisburg, or Richmond—or emerge from numerous sources in the federal government, including the Environmental Protection Agency (EPA), Congress, the White House, and the U.S. Supreme Court. As if that were not enough, the people who have impacted this story range from grassroots volunteers to U.S. presidents. This cacophony of notes from the past can be turned into the symphony of history by the Chesapeake Bay itself, which is why we will begin by taking a bird's-eye view of the estuary and the challenges it faces.

To better appreciate one of the key themes of the Bay's recent human history, it is helpful to imagine the Bay as an osprey might see it. Imagine an osprey flying north from her winter retreat in Central America in search of a good location to hunt, nest, and breed in the Mid-Atlantic. As she flies north from the Bay's mouth at Norfolk, Virginia, to its head at Havre de Grace, Maryland, and beyond across Pennsylvania and up to the headwaters of the Susquehanna—the Chesapeake's main tributary and source of roughly half the freshwater entering the Bay—near Cooperstown, New York, America's largest estuary unfolds as a single habitat. She is ignorant of the political divisions that crisscross the 64,000-square-mile watershed, the 11,000 miles of shoreline, and the 200-mile main stem of the Bay. During the spring

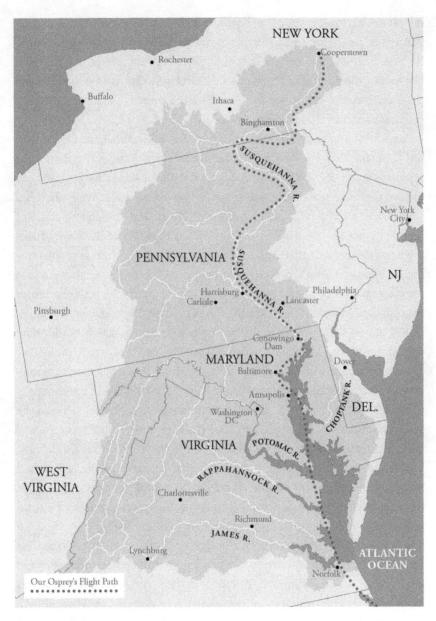

The Chesapeake Bay Watershed. (Illustration by Turner Blashford, from USGS base map)

migration, an osprey can travel a hundred miles or more in a day, meaning that it would take her about five days to fly from one end of the watershed to the other. Although to her it is a single expanse of water, wetland, and wood, historically the Bay has not been treated as one unit. On her first day going up the coastline, midway up the Bay she will reach perhaps the most important political boundary, the Potomac River.

Residents and state officials of Maryland and Virginia have fought over the Potomac River's resources and management since the colonial era. The problems go all the way back to the original charter King Charles I granted to Lord Calvert, the Baron of Baltimore, in 1632. According to King Charles I, a massive tract of land, which became the colony (and subsequent state) of Maryland, would have a border from "the first Fountain of the River of Pattowmack . . . unto the further Bank of the said River, and following the same on the West and South, unto a certain place . . . where it disembogues into the aforesaid Bay of Chesapeake." In plain English, Lord Calvert's charter for the colony of Maryland included all of the Potomac River (the *further* bank), rather than splitting it down the middle with Virginia, which would more commonly be the case with aquatic borders. This has been a sore subject for generations, up to and including a 2003 U.S. Supreme Court case, *Commonwealth of Virginia v. State of Maryland*, in which Maryland tried unsuccessfully to block Virginia from using the river. The court eventually ruled 7–2 in the commonwealth's favor, concluding that the Maryland-Virginia Compact of 1785 had granted Virginia certain riparian rights to the Potomac. This aquatic border has been contentious in no small part because the broad juncture where the mouth of the Potomac meets the Chesapeake Bay might be a great place for our osprey to nest for a night but a terrible place to enforce political jurisdiction. Indeed, the history of conflict is so recent that our osprey could find herself resting for the night in a loblolly pine that bore witness to the last fatality of the Oyster Wars, in 1959, when Maryland's Oyster Police shot and killed a Virginia waterman who they thought was fishing on the wrong side of an invisible line extending from the mouth of the Potomac out into the Bay. Although ancient history to current residents of Maryland and Virginia, in the 1960s memories of violence and conflict between the two states over the Bay were recent and raw, and a serious impediment to cooperation.[4]

As our osprey continues flying north, on her second day she will pass over Baltimore and Havre de Grace, reaching the mouth of the Susquehanna

River, the Bay's largest tributary, and by extension its greatest source of pollution. If our osprey is feeling motivated, she can easily make it to Conowingo Dam by nightfall. Immediately behind Conowingo lie nearly three hundred million tons of sediment. Conowingo's sediment tells the story of the Bay's water quality issues in a nutshell. Trapped in that sediment are millions of pounds of nitrogen and phosphorous that have washed down the Susquehanna from major sources like farm fields, suburban development, and wastewater since the dam's completion in 1929. The nutrients and sediment that are not trapped behind Conowingo pollute the Bay by smothering underwater grass beds and oyster reefs, clouding the water to inhibit the growth of underwater vegetation, and acting as a fertilizer for plant plankton, which sets off a harmful chemical reaction known as eutrophication, which ultimately results in large "dead zones" where the Bay's waters are too deprived of oxygen to support most animal life. If our osprey finds herself fishing in a dead zone, she and her chicks will be out of luck. Best to keep flying north.[5]

The Susquehanna River itself is nearly 450 miles long, so it would take our osprey a few more days to fly all the way to the river's headwaters. Along the way, she might fly over Harrisburg, Pennsylvania's capital, or perhaps over Lancaster, its agricultural hub. One of the most intractable problems Chesapeake environmentalists had to confront was the fact that although Pennsylvania accounts for the largest chunk of the Chesapeake's watershed, no part of the Bay extends north to the Keystone State. It was not until the 1970s that scientists began to realize how severely agricultural pollution from Pennsylvania harmed the Bay's water quality. Although many individual residents of Pennsylvania benefit from the Bay's seafood and recreational opportunities, much of the economic gain goes to Maryland and Virginia. Persuading legislators in Harrisburg to coordinate with their counterparts in Annapolis and Richmond has been a massive challenge, but since the mid-1980s Pennsylvania has been a member of the federal-state Chesapeake Bay Program and the tristate Chesapeake Bay Commission. Our osprey is ignorant of all this, but what she does know is that there is excellent trout fishing in some of Pennsylvania's streams, such as the Letort Spring Run, which flows through Carlisle and joins the Conodoguinet Creek, which curves sinuously east for a few more miles before emptying into the Susquehanna River just north of Harrisburg. Chesapeake environmentalists made a similar observation and worked to forge a connection to the Bay through the streams and creeks that run through many Pennsylvanians' backyards or local parks.

As our osprey completes her journey north and finds a suitable spot for fishing and raising her chicks, she has entered one of the most recent additions to the Chesapeake restoration effort, New York. In 2014 the Empire State, along with West Virginia and Delaware, signed the Chesapeake Watershed Agreement, which for the first time brought all six watershed states, plus the U.S. government (via the EPA) and Washington, DC, into a common pact to create "an environmentally and economically sustainable Chesapeake Bay watershed with clean water, abundant life, conserved lands and access to the water, a vibrant cultural heritage and a diversity of engaged stakeholders." Although the three newcomers do not feature prominently in the history that follows, their inclusion in the 2014 agreement reflected an important milestone in the evolution of the Chesapeake Bay from a battleground between Maryland and Virginia to a watershed-wide restoration initiative led by the federal government. Like our osprey, many humans who encounter the Bay and its tributaries might be more interested in its scenic shoreline and tasty seafood than in the arcane web of laws, agreements, and policies that govern the Bay, but one piece of the policy framework in particular, a Total Maximum Daily Load (TMDL), is worth understanding in more detail before moving on.[6]

Just as our osprey could complete a journey from the Bay's mouth to its headwaters, so too could oceangoing fish like shad complete voyages deep into their natal streams. These and other species are part of a natural cycle of nutrients that makes up the organic machinery flowing in and out of the estuary and its watershed. Human action has significantly disrupted this nutrient flow, and whatever their origins, once in the Chesapeake, excessive nitrogen and phosphorus cause significant harm to the ecosystem. Yet it was not until 2010 that the EPA began to enforce a limit on nutrient pollution, derived from decades of scientific research, which treated the Chesapeake, in theory at least, as a single ecological unit. Operating under a consent decree, the EPA established a Total Maximum Daily Load of nitrogen, phosphorus, and sediment pollution. Nicknamed the Bay's "pollution diet," the TMDL set an enforceable maximum acceptable limit of pollution entering the Bay derived from scientifically determined limits on the different tributaries and subwatersheds constituting the ecosystem. It was the largest TMDL created by the EPA, and in 2016 it survived a legal challenge that went to the U.S. Supreme Court. The TMDL is one of the key guides to restoring the Bay, and it reflects both the potential and the shortcomings of the movement. The watershed states have made fitful progress restricting the "calories" of pollution entering the Bay, but the region's political leaders acknowledged in 2022 that they would fail to meet a 2025 deadline to have practices in place to reduce levels of pollution to the EPA's target. Although the Chesapeake Bay restoration effort encompasses much more than three types of pollution, the story of nutrient pollution and the TMDL has played a major role in the development of a federal leadership role in the Chesapeake's recovery. While many important restoration projects, such as protecting commercially valuable species like striped bass, crabs, and oysters, saving valuable wetlands and forests, and even some forays into nuclear power and oil refining, feature in this account, the dominant question during the first fifty years of the Bay's restoration has been what to do about nutrient pollution from the individual states in the watershed, and consequently that is one of the main themes of this book.[7]

HISTORICAL ACTORS AND CHRONOLOGY

If the major theme of this work can be summarized as the development of an EPA-led recovery strategy that treats the Bay and its major pollutants as a single ecosystem, the major historical actors form anything but a single coherent unit. In broad strokes, concerned citizens, elected and appointed officials at the state and federal levels, scientists, commercial and recreational fishers, farmers, landowners, and, yes, professional environmentalists all pushed the Bay's environmental politics in the direction of a unified approach, even if they did so for sometimes conflicting reasons and in pursuit of sometimes contradictory goals. Still, despite perennial disagreement over the how and especially who will pay, these groups have maintained a dogged dedication to the common goal of protecting and restoring the Chesapeake Bay. Much of the credit for the movement's consistent focus can be attributed to the Bay itself, which looms so large in the region's culture, economics, and physical landscape that ignoring its plights would be all but impossible. As we will see, however, many of these same groups have lamented harmful changes in the Bay since at least the late nineteenth century, which indicates that something different happened in the 1960s and 1970s to galvanize a movement.

The rise of the Chesapeake Bay Foundation (CBF), the regional movement's oldest, largest, and most influential Bay-focused environmental organization, undoubtedly played a part. Officially chartered in 1966, CBF is one of the most consistent actors in this book, though the fact that CBF shows up often in this history should not cause one to mistake this for a history of CBF. The genius of CBF's early founders was not that they created an environmental group—many environmental groups were emerging in the ferment of the late 1960s—but that they envisioned their group's primary purpose to be to advocate for all of the Chesapeake Bay at a time when few other individuals thought of the Bay as an ecological whole. For instance, one of CBF's cofounders, the Pulitzer Prize–winning journalist Felix Morley, complained in 1966 that in Chesapeake affairs "there were a lot of specific end purposes and no coordinating agency," which left him "wondering why there had been no 'Chesapeake Foundation' before." While not the sole actor, and at times overshadowed by other actors and events, CBF is a significant throughline in this history, perhaps second only to the Bay itself. Therefore, chapter 1 begins by examining the rise of CBF and its role in the emergence of a new regional environmental movement. This account takes advantage of never-before-used archival sources, oral histories, and internal documents to examine the birth of the Chesapeake Bay Foundation in the mid-1960s, and its relationship to the rest of the emerging Chesapeake Bay environmental movement in the years leading up to the first Earth Day in 1970. Although numerically small, CBF and its allies foreshadowed future trends in the battle over the Calvert Cliffs nuclear power plant, which led to significant changes in federal environmental and nuclear policy. However, CBF did not succeed during its first decade of existence in creating lasting coordination of efforts to protect the Chesapeake Bay.[8]

Given the EPA's current high level of involvement with the Bay, one of the most shocking facts about the Chesapeake's environmental history is that the agency had a minor, almost negligible role in the Bay's environmental protection until the mid-1970s. As a leading environmentalist lamented, "Here on the Chesapeake, we tended to doze."[9] The EPA would not seriously get involved in Chesapeake issues until 1975, with the creation of the Chesapeake Bay Program, which began as an EPA research portfolio before evolving into a joint federal and state restoration initiative in 1983. The wake-up call for Chesapeake environmentalists is the main storyline of chapter 2, which analyzes how scientists, commercial fishers, known as *watermen* in the region, and environmentalists responded after Tropical Storm Agnes to agitate for a more coordinated approach to understanding, managing, and protecting the Bay. With the help of a powerful patron in Congress, Senator Charles Mathias (R-MD), these groups successfully lobbied for an EPA study program that would explain the causes of the Bay's decline and point to a pathway to restoration.

Chapter 3 continues the narrative of the EPA's involvement in the Bay but shifts from the realm of science and resource use to the political arena, following the story of the Chesapeake's restoration all the way to the office of President Ronald Reagan. Although famously opposed to big government and environmental regulations in general, and skeptical of the Chesapeake in particular, Reagan ultimately backtracked after confronting a powerful pro-Bay, bipartisan environmental coalition. In a moment of dramatic historical reversal, it was under Reagan's administration that the major expansion of the federal government's role in Chesapeake affairs began. Sensing a winning cause, Reagan even highlighted the Bay in his 1984 State of the Union address, arguing in words that would shock the sensibilities of twenty-first-century conservatives that "preservation of our environment is not a liberal or conservative challenge, it's common sense."[10] Chapter 3 covers the shortest time period chronologically, but the play-by-play account of Chesapeake environmental politics during Reagan's first term reveals major insights into American environmental policy and the changing nature of the Republican Party, themes that are further developed in chapter 5.

Whereas chapter 3 follows the politics of Bay restoration in the halls of power in Washington, DC, chapter 4 explores and analyzes the first major wave of Chesapeake policy achievements in Maryland, Virginia, and Pennsylvania, from the mid-1980s to the early 1990s, which stopped the Bay's decline and began nudging it tentatively on the road to recovery. In 1985 the *Washington Post* editorialized that "people may not be 'into environmentalism' the way they were a few greener years ago, but the most refreshing revival of pure and simple concern about mucking up our surroundings is popular support for saving the Chesapeake Bay."[11] This history complicates and challenges common assumptions about the 1970s being the "environmental decade" and the 1980s being a period of rising conservative backlash. At least in the Chesapeake region, the story is more tangled, but by the early 1990s trends in the region converged with national trends pushing back against the gains of the environmental movement. For Chesapeake environmentalists, what began as earnest optimism following the first Chesapeake Bay Agreement in 1983 evolved over a decade into steely resolution to continue the struggle despite mounting opposition, antagonism, and apathy.

Chapter 5 begins at the nadir of the Chesapeake Bay environmental movement, in the early to mid-1990s. Despite meaningful accomplishments in the prior decade, public attention for the Bay waned, and a new strain of antienvironmental thought infected conservative Republicans, who sought to undo much of the Chesapeake Bay Program's work after they scored major electoral victories at the state and federal levels. Yet, bipartisan support for the Bay at the local level remained remarkably resilient, and when given a chance by another environmental disaster in 1997, the movement surged back to life. The spark was unexplained fish kills in the estuary that were blamed on the mysterious toxic microorganism *Pfiesteria piscicida*. The millions of dollars in lost revenue to the seafood and tourism industries underscored the economic value of a healthy Bay. Following the *Pfiesteria* disaster, Chesapeake environmentalists built momentum for a winning streak of new policies and funding at the state level. Although bipartisan enthusiasm for the Bay remained high in the region, Chesapeake environmentalists were unable to translate the Bay's popularity into success at the federal level. Repeated setbacks in legislative and executive policy formation led to the climax of this period in the Chesapeake's environmental history: a lawsuit by CBF and other allies that resulted in the EPA formulating the Chesapeake's Total Maximum Daily Load of nitrogen, phosphorus, and sediments under a consent decree. Although there have been important developments since the EPA established the Chesapeake TMDL in 2010, there is great risk for historians who stray too close to the present day, and this narrative ends with the TMDL clearing its final legal challenges and heralding a new era of federal involvement in Chesapeake environmental history.

THE TEST OF THE CHESAPEAKE

The Chesapeake Bay's greatest biologist of the twentieth century, and quite possibly its most politically influential scientist, was Eugene Cronin, who got his start studying blue crabs before leading the University of Maryland's Chesapeake Biological Laboratory. He served on numerous scientific advisory boards and commissions and for a quarter century was a CBF trustee. He was a crucial bridge between the worlds of science, policy, and advocacy. In 1970, on the eve of the first Earth Day, he wrote a short article arguing that "our estuaries offer a compelling test of our ability to live in enduring harmony with our environment." Cronin detailed the value of estuaries for humans as well as for the flora and fauna that can thrive in the unique mixture of brackish water, tides, river currents, dry lands, and wetlands that form the estuarine ecosystem. Cronin lamented that "we do not yet sufficiently understand estuarine systems or processes" and that many estuaries were "vivid examples of environmental insult—the result of man's failure to achieve a durable relationship with his environment." At the time, Cronin did not number the Chesapeake among the vivid failures, but he did warn his fellow scientists that they "must take a more active part in comprehending the nature and capacities of such ecosystems as estuaries and sounding a clear klaxon when they are threatened." Cronin would go on to do exactly that for the next three decades of his career, playing a major role in advancing the understanding and protection of the Bay ecosystem.[12]

A generation later, William Cronon, a leading environmental historian, took up the challenge of interpreting five hundred years of Chesapeake environmental history in his 2001 essay "Reading the Palimpsest." Like Eugene Cronin, Cronon found grand implications for humanity "in contemplating the dilemmas we now face in managing the Chesapeake." Cronon argued that "by taking responsibility for the extraordinarily complicated relationships that make up an ecosystem like the Chesapeake, we open ourselves to the knowledge and wisdom that can lead to an ever deepening appreciation for the complexity of the world we live in." Ultimately, according to Cronon, the lessons from the Bay's long history are "our only available guides for charting our way into the uncertain future." Beyond the charming homophonic quality of their surnames and their titanic stature in their respective disciplines,

Cronin and Conon also shared the view that the Chesapeake in particular, and estuaries in general, offered an existential challenge for the human species. With the advantage of being able to analyze subsequent decades of history since their writing, this book advances the general project of uniting history and science to assess how well humans have met this challenge. Indeed, considering the work of those two eminent scholars, might we not ask the logical follow-up question: If estuaries are places where the relationship between humans and the natural world faces the ultimate test, what better place to seek knowledge and wisdom about how to meet those challenges than in the example of those who tried to pass the test of the Chesapeake?[13]

Before giving the historian's answer to how well Cronin and his counterparts fared, it is worth reviewing what more contemporary scientists have to say about the Bay. Fortunately, a deep dive into peer-reviewed literature on estuarine hypoxia, benthic zones, or thermal dynamics of the pycnocline is not necessary to form an assessment of the Bay's water quality, because there has been a proliferation of report cards on the health of the Chesapeake Bay. Although imperfect, these form a valuable starting point for assessing how well the Bay is doing. Two of the more high-profile Chesapeake Bay report cards, one by CBF, the other by the University of Maryland Center for Environmental Science (UMCES), show modest gains for the Bay since the beginning of the restoration effort. In 2021 UMCES gave the Bay a C grade and reported that despite mixed results, "the positive news is that the overall Bay trend is still improving." In 2022, CBF gave the Bay a D+ and emphasized that while there was progress to celebrate, "the Chesapeake Bay, and the thousands of rivers and streams that fed it, are not as healthy as they can—and should—be." Thus, as of this writing it is not clear whether the effort has been sufficient to pass the Chesapeake's test with flying colors; at the same time, it is abundantly clear that it has not been a failure.[14]

As the Chesapeake Bay restoration project reaches its 2025 deadline for water quality with its goals once again unmet, questions rightly abound regarding the future of the estuary, what recovery goals should be pursued, and how to best achieve them. At this moment, a historical perspective is vital to remind us that this is not the first time people fearing for the Bay's future have asked these questions and found the resolve to see them answered. Klaxons are sounding, as they have before, and despite much progress, the future of the Bay remains uncertain. Most of the actors who dominate the subsequent pages, the stars of the drama of the first fifty years of the Chesapeake Bay's environmental history, will not have much of a role, if any, in what happens next. That will depend on a new generation of Chesapeake environmentalists, who will have the formidable task of answering those questions and thinking of the new ones that need to be asked. It is to those who will carry on the work that this book is dedicated, in the hopes that the history in these pages will serve as a useful study guide for all who wish to ensure that we pass the test of the Chesapeake.

ONE
"SAVE THE BAY"

Identifying where the Chesapeake Bay begins is relatively easy. Consult a United States Geological Survey fact sheet, and you will see that the Chesapeake Bay begins at the freshwater mouth of the Susquehanna River, at Havre de Grace, Maryland, and ends just past Virginia Beach, Virginia, where it empties into the salt water of the Atlantic Ocean. From fresh to salt water, with a brackish mixture in between. This is the archetype of an estuary. But, if you want to suggest that that the Chesapeake Bay is really just the tidal portion of the Susquehanna River, that is not a problem. The USGS has helpfully identified the Susquehanna's source at Otsego Lake, near Cooperstown, New York. If you want to go even further and ask, "Well, where does *that* water come from?," you will find a number of small streams flowing into the lake, for each of which a precise set of coordinates indicates its source. Beyond that, it is either rain falling from the sky or groundwater swelling up from beneath the soil. Repeat this exercise with every major (or minor) tributary of the Chesapeake Bay, and you will find yourself tracing the outlines of a 64,000-mile, six-state watershed. That is relatively easy. On the other hand, protecting and restoring the nation's largest estuary and its accompanying wetlands and forests, underwater grasses and fishes, oysters and crabs—in sum, the biotic components encircled by that watershed outline—is a challenge.[1]

The ongoing effort to protect and restore the Chesapeake Bay has been driven by a regional environmental movement that this book calls *Chesapeake environmentalism*. Chesapeake environmentalism is the combined work of hundreds of nonprofit organizations, state and federal agencies, research scientists, elected officials, and concerned citizens. In the past decade, Chesapeake environmentalism has leveraged an average of $1.5 billion annually in state and federal funding to protect and restore the Bay. It is a story of political and ecological success, albeit an incomplete, messy, work-in-progress type of success. One cannot identify a single moment when it "began"; there is no clear chronological analog to the geography of the Chesapeake's watershed, where east of a line through the Allegheny Mountains water will flow to the Bay and to the west it will eventually wind its way past Pittsburgh and on down to the Gulf of Mexico via the Ohio and Mississippi Rivers. That said, one could do a lot worse than starting with the region's oldest, biggest, and historically most influential environmental group, the Chesapeake Bay Foundation.

CBF and the history of Chesapeake environmentalism matter beyond the watershed. They exemplify a regional environmentalism that points toward a richer understanding of environmental movements in the United States. One of the central arguments of this book is that what is often referred to as "U.S. environmentalism" or "the U.S. environmental movement" is not a monolithic social movement but a complex, multifaceted set of individuals, groups, and institutions that sometimes work together but other times work at cross-purposes. This in itself is not a radical argument; scholars have recognized that environmentalism is often "a fragmented movement" in which "larger national groups work inside the Beltway, while the grassroots organizations remain ad hoc, single-issue organizations" without much broader influence. What is new here is that the example of Chesapeake environmentalism smashes this false dichotomy of larger power players versus single-issue local groups by showing how different groups of people and organizations fit together to create a movement that has had a powerful impact on federal policy while remaining grounded in a specific place. This book argues that the movement's remaining grounded in the specific ecological contexts of the Chesapeake is one of the reasons why it has sometimes been able to overcome the increasingly toxic partisanship of U.S. politics. This study suggests that further research into the reciprocal relationship between ecology and politics at both the state and the federal level of government will yield a more complex and nuanced understanding of U.S. environmental movements.[2]

THE BIRTH OF THE CHESAPEAKE BAY FOUNDATION

Before there was a Chesapeake Bay Agreement, before the U.S. Environmental Protection Agency came into existence, before even the first Earth Day, there was the Chesapeake Bay Foundation. Officially incorporated as a 501(c)3 nonprofit in 1966, it quickly became known as the "Save the Bay foundation" because of the iconic bumper stickers that quickly came to symbolize the movement. Unlike environmental groups that grew out of suburban concerns, CBF originated in the social circles of yacht club members and gentlemen farmers in Annapolis and Baltimore. Understanding CBF's origins is especially important to understanding Chesapeake environmentalism as a whole because CBF played a significant role in shaping that movement. CBF's beginnings also illustrate the power of a specific place, in this case the Chesapeake Bay, to inspire an environmental movement that reflects the particular ecological and cultural components the movement sought to protect and defend. CBF would change significantly over the decades, but during its earliest years it proved capable of linking local and national environmental movements through a range of issues, with none more important than nuclear power.

Original "Save the Bay" bumper sticker, 1969. The Chesapeake Bay Foundation in particular and the movement in general would soon become synonymous with this slogan coined by Porter Hopkins. (Photo by author, Special Collections and University Archives, University of Maryland Libraries)

The Chesapeake Bay Foundation grew out of a social network of upper-class men who enjoyed recreational sailing on the Bay. CBF's founding cohort clustered around two foci in the region: the Gibson Island Club, located on a small private island just north of Annapolis, on the other side of the Chesapeake Bay Bridge, and the Windjammers, a group of yacht enthusiasts that sponsored recreational and educational events related to the Chesapeake's sailing heritage. These two foci were linked through Arthur Sherwood, CBF's first president and one of the most influential leaders in CBF's early years. CBF's initial promotional materials described the group as "growing from a nucleus of a few Windjammer members." Although a spinoff of the Windjammers, CBF would not linger in its parent's shadow for long. While on the one hand CBF's origins set the group apart, because its founders predominantly viewed things from the point of view of recreational sailors, uncommon in environmental groups; on the other hand CBF's origins were similar to those of the many environmental movements that sprang from recreational uses of nature, such as hunting, mountaineering, hiking in wilderness, or even driving. This matters because while these men—and they were all men—were privileged and saw the uses of the Bay in terms of their own recreation, they did possess an intimate, firsthand knowledge of the Bay. To them the Bay was not an idealized abstraction nor a pristine Eden to be sheltered from the hand of man. It was their Chesapeake Bay. They knew it well, they loved it, and by the early 1960s it was clear to them that something was wrong.[3]

C. A. Porter Hopkins, one of the last living members of CBF's founding cohort, is a case in point. In a 2012 interview with the author at his farm outside Cambridge, Maryland, he explained that abrupt changes in the ways he interacted with the Bay, such as in hunting waterfowl, had made it clear to him that something was wrong. Hopkins recalled, "I came at it from the standpoint of seeing the things that I really loved to do, not being able to do them anymore, because they weren't there." He was, by his own admission, a bit of a black sheep, more interested in uses and abuses of the land than many others in his social circle. He was from a family of bankers, but he had fallen in love with the land and chosen to become a farmer after serving in the U.S. Army during World War II. Hopkins's entry point to conservation was the rapid destruction of the Chesapeake's marshes, which he saw vanishing quickly in Maryland's postwar economic boom. "Well, I love wetlands," Hopkins recalled, noting that "people here [in Maryland] finally recognized that we ought to start thinking about not filling all our damn wetlands up. Because this is where the ducks want to be, this is where the muskrat are, this is where the oysters, the crabs, and the fish are all coming from." As Hopkins saw the rapid development of Maryland's farmland and wetlands, he got involved in politics and what we might today call "community organizing" to do something about it. Hopkins helped start several conservation groups in Maryland, and as a lifelong Republican he served in the Maryland General Assembly from 1967 to 1978, first as a state delegate, attaining the rank of minority leader, and then in the state senate, before retiring to his farm on Maryland's Eastern Shore. Hopkins was an unconventional legislator—he recalled that "everybody said, 'this guy Hopkins is sorta off the wall'"—but it took some unconventional thinking to see the need for the Chesapeake Bay Foundation.[4]

Creating a Chesapeake Bay foundation was not a necessary or obvious step in the 1960s. In the first place, there were already many organizations working on some aspect of the region's environment or culture. At one of their early planning meetings, in May 1966, before officially incorporating as CBF, the founding group drew up a list of twenty-five private and public agencies and groups whose work was connected with the Bay. They would later enlarge that list to include more than fifty groups, among them CBF's parent organization, the Windjammers. (The Chesapeake Bay Program website now lists more than six hundred groups.) Indeed, one of the founding group's first debates was over whether CBF should be a separate entity or just a Windjammer project. One forceful early proponent of a Chesapeake Bay foundation was Arthur Sherwood, a Baltimore attorney, avid sailor, and on-again, off-again Republican candidate for mayor of Baltimore. Sherwood was convinced that without an independent group to represent the Bay as a whole, the many different groups using the Bay or concerned with various aspects of the Bay would tear it apart. Sherwood's widow, Suzanne, recalled that when Sherwood pitched his idea of a Bay-wide citizen interest group to his friends, "most of them thought he was crazy of course."[5]

One person who thought Sherwood was on to something was none other than Porter Hopkins. Hopkins knew many of the men who made up CBF's founding group through his connection to the Gibson Island Club. In particular, he had grown up in the same social circles as Arthur Sherwood (and briefly entertained the idea of dating Sherwood's younger sister). Hopkins thought Sherwood was "bright as hell" and credits him with being "the driving force" behind CBF's creation. However, Hopkins was also quick to point out that at the same time that he was getting to know Sherwood's friends in the Windjammers, many other groups were beginning to take interest in the Bay, including chapters of national organizations like the Izaak Walton League and the Audubon Society, as well as strictly local groups like the League of Maryland Sportsmen. According to Hopkins, concern for the Bay was rising, and "the tone was there and the table was set" for "a major conservation move on the part of the Chesapeake Bay." It was Hopkins's sense of urgency and strong political instincts that later led him to coin the phrase *Save the Bay* and put it on bumper stickers that people could buy to raise money for the group. Archival evidence confirms the pivotal roles that Sherwood and Hopkins played in the inception of CBF. At a June 1966 planning meeting, it was Sherwood who first suggested "a definite resolution not to associate ourselves" with the Windjammers and it was Hopkins who called for CBF to act as "a 'watchdog organization,' a non-political group acting for the citizens." With the need for the organization and its mission loosely established, Sherwood filed the paperwork with the IRS for tax-exempt status. On November 23, 1966, CBF was born.[6]

LAYING A STRONG FOUNDATION

Despite the promising group of well-connected elites that comprised its board and the favorable local and national conditions for environmentalism in 1966, CBF's founders had a hard time getting the organization off the ground. Reading their meeting minutes and correspondence from 1966, one is tempted to conclude that this was a group of wealthy dilettantes who expected donors to emerge like mushrooms after a spring rain. Although CBF's founding cohort of fourteen men included four company presidents, a congressman, the governor of Maryland, and a Pulitzer Prize–winning author, none had any experience in running an advocacy organization (or, in their defense, the time to do it). They did, however, have abundant optimism about their new group. One of CBF's two vice presidents, Marshall Duer, expected CBF to quickly become "a coordinating group around which the many other civic and anti-pollution and improvement organizations can rally." Richard

Randall, CBF's other vice president, concurred, writing that "by the time we send a few bulletins around, we'll get so many ideas, we'll need a staff bigger than any million dollars can supply." Arthur Sherwood, newly minted as CBF's first president, was only a bit less sanguine and set a modest five-year fundraising goal of two hundred thousand dollars.[7]

Although CBF's founders were optimistic, the organization was nearly dead on arrival. By November 1967, a full year after receiving tax-exempt status, CBF had only raised $5,200. However, CBF's founders had scraped together just enough to hire the group's first employee, Jess Malcolm, and rent an office across the street from Maryland's capitol building. Although not a founder in the literal sense of the word, Malcolm shaped CBF's early development more than anyone else and along with Arthur Sherwood ensured that the group would survive past its rocky first years. Malcolm and Sherwood would come to personify two competing visions for CBF, and more broadly, they represented two different approaches to environmentalism. A biologist, Jess Malcolm was an aggressive purist who believed that threats to the Bay's health must be stopped at all costs, and he saw the ecological integrity of the Bay as CBF's most important goal. He considered himself a "biological philosopher," and he saw policy and law as the primary weapons for achieving this goal. Sherwood was a more conciliatory peacemaker who hoped to find a balance between the competing uses of the Bay by rejecting what he called "militant environmentalism" and instead building a movement based on "environmental reasonableness." At the core of Sherwood's environmental philosophy was his view that the competing uses of the Bay could be harmonized, although he was sometimes at odds with the scientific reality. The tension between the two men would ultimately become destructive and lead to a power struggle atop CBF; however, at first their tension was dynamic, jolting CBF out of the neutral gear in which it had been stuck during its first year in existence.[8]

Sherwood and Malcolm both recognized that while they understood that the Chesapeake region needed a group that would represent the Chesapeake as a whole, few others agreed. Indeed, one of the biggest challenges CBF faced throughout its earliest years was finding acceptance outside Maryland. One of CBF's first trustees from Virginia put it bluntly: "CBF was perceived as a Maryland organization and that was a problem." One of CBF's early fundraising strategies was to turn to the numerous groups concerned with some aspect of the Bay in Maryland and Virginia and ask them to support CBF's overall efforts on behalf of the estuary. This did not go well. In a message to the board, Sherwood unloaded on other groups and users in the region. Calling it an "aggravating effort," Sherwood complained of "the ghastly reluctance of the very users of the Bay to put just a token dollar where their hearts seem to be." To Sherwood and many on the board, CBF's value was self-evident. Malcolm, however, acknowledged that "for the most part, we had yet to prove our worth."[9]

Fortunately for CBF, Sherwood was able to draw upon his connections through the Gibson Island Club to buy Malcolm some time to prove CBF's worth. One of Sherwood's childhood friends was C. Trowbridge "Tobe" Strong, a scion of the family that cofounded Eastman Kodak. Tobe had been raised in New York, but his father was a passionate sailor and member of the Gibson Island Club. Strong had spent his summers growing up with Sherwood, and the two had become close friends. Strong had been a groomsman at Sherwood's wedding, and later Sherwood would write that "when I think of friendship, I think of Tobe." Unsurprisingly, with CBF's finances in dire straits, Sherwood turned to Strong for help. Sherwood was still seeking a commitment of two hundred thousand dollars over five years to get CBF off the ground. Although they were close friends, this was a little too much at first. However, Strong persuaded his brother and his mother to use their family's philanthropic foundation to give CBF a chance with an initial two-year grant of twenty thousand dollars. Crucially, Strong believed he could get Sherwood the money he wanted if CBF could prove it would be a worthwhile investment in memory of Strong's late father, L. Corrin Strong. As Sherwood put it to CBF's board, "We can be encouraged at the prospect of immediate financial help for a trial (one or two year) period; challenged by the prospect of a substantial endowment in memory of L. Corrin Strong if it appears we deserve it."[10]

Throughout much of 1968, CBF did not appear to deserve it. The group struggled to raise money even from its own board of trustees, and it seemed incapable of making an impact that generated any sort of publicity. Without the initial support of the Strong Foundation, CBF likely would have been defunct by the end of the year or at best served as a glorified social club in which local elites could complain about the Bay's decline. Sherwood was anxious to put the Strong money to good use, and he worried that despite his connections, his fellow trustees' lackadaisical attitude toward fundraising might squander CBF's opportunity. At the foundation's February 1968 board meeting, he urged the rest of the board "to give more thought to the problem of fund-raising" and warned them that the ten thousand dollars for 1969 was contingent "on the success of our efforts in 1968." Despite repeated pleas, most CBF board members took a hands-off approach to fundraising, and the organization lacked any concrete results to justify its existence. By May 2, even after mailing twelve hundred copies of a "Chesapeake Bulletin" as part of its first membership drive, CBF had a grand total of thirty-four members. More damning than that, it had raised only $375 since receiving the Strong donation. With Sherwood having played his hand with his social connections, it was left to Jess Malcolm to run the day-to-day affairs of the organization and make an impact worthy of the memory and money of L. Corrin Strong.[11]

MALCOLM'S MOMENT

In August 1968, CBF got its opportunity to make a splash when Jess Malcolm learned of a brewing grassroots opposition to a planned oil refinery at Piney Point, Maryland, in rural St. Mary's County near where the Potomac River meets the Chesapeake (see map on p. 34). Local citizens' groups were already opposed to the planned refinery but had not generated broader awareness of the potential threats it posed. CBF played an important though not singular role in the effort to stop it. While the Piney Point oil refinery helped galvanize an embryonic environmental movement in the region, the long-term effects on CBF's subsequent ability to advocate for the Bay would prove to be consequential. In the short term, while CBF scored its first victory, Malcolm's tactics caused controversy within the organization and exposed fault lines that would recur throughout CBF's history and the history of Chesapeake environmentalism.

Malcolm first learned of plans by the Steuart Petroleum Company to build an oil refinery near the mouth of the Potomac when he received a phone call from a worried citizen in St. Mary's County who wondered what CBF could do to help publicize their plight. Seizing the opportunity to get involved, Malcolm sent out an inflammatory press release that helped stir local newspaper coverage of the story. He intuitively picked an issue that would appeal to many in the region: waterfowl. Malcolm claimed that the oil refinery would threaten 50–70 percent of the duck population in the Atlantic flyway, and he accused the oil company and government officials of acting in a "surreptitious manner." There had been no public discussion of the project (nor was any legally required at that time), and Malcolm castigated state officials for virtually approving Steuart's proposal "without public hearings or any other attempt to inform local residents of the project." Malcolm's scathing statement that "such secrecy is totally unacceptable under our form of representative government" was a juicy soundbite, especially coming from an organization that counted prominent local, state, and national politicians among its trustees and founders. Malcolm and his allies in St. Mary's County succeeded in turning a perfunctory permitting issue into a high-profile policy struggle.[12]

Malcolm's comments drew predictable criticism from industry sources, but perhaps to his surprise, he also received heavy criticism from within CBF. Donald Pritchard, the director of the Chesapeake Bay Institute at Johns Hopkins, was one of the region's most respected scientists. He had been invited to join CBF's board in February 1968, and he was not happy with Malcolm's comments. Pritchard was an influential trustee despite being relatively new to the foundation; he was the chair of CBF's Policy and Scientific Review Committee, and his words carried weight because of his stature as one of the most eminent scientists in the region. He wrote to CBF's board that after reading Malcolm's press release, "I must admit that I was somewhat taken aback by certain aspects of the news release. My main concern is that it tended to imply that Maryland State agency personnel and other officials of the State were at least negligent if not biased against the public interest in this matter." Further, Pritchard added that in terms of long-term strategy he was "concerned that the Foundation will not become an effective voice if we alienate these agencies," which he said was precisely what Malcolm risked doing "by suggesting improper action on the part of a public agency."[13]

Pritchard directed most of his ire toward Malcolm, but with the keen eye of an outsider new to an organization, he also called out CBF's board for not having a clear vision for CBF. He even excused Malcolm somewhat, on the grounds that the board "did not give very explicit guidelines to Jess" about how he should execute CBF's loosely defined mission. Pritchard called for the board to "very soon arrive at some definite decision as to its ultimate goals." He warned that by simply taking stances against development of the

Bay, CBF could appear to be a successful organization, but he added that unless CBF developed a specific goal rather than a scattershot opposition to major projects, he "doubt[ed] that the Foundation can be really effective in the long run." Without definitive guidelines from the board, Malcolm had gone ahead and pursued the matter as he saw fit. Unlike Pritchard and others on the board who wanted to maintain a cordial relationship with industry and government contacts, Malcolm thought the way to achieve the foundation's goals was by mobilizing the citizenry for a head-on confrontation against threats to the Bay. Malcolm put it bluntly to the board in his 1968 annual report, telling them that "the war for Chesapeake Bay is being lost." The way to save the Bay, according to Malcolm, was for CBF to develop "a new philosophy" that would "assure its citizens that short term economic quality gains and Bay quality sacrifices are no longer acceptable." Malcolm was unquestionably a hard-liner. To him, anything "involving unknown or detrimental effects to the Bay should be denied unequivocally." This stance unnerved some within CBF, including Sherwood and Hopkins (who in particular was unimpressed), but for the time being, Malcolm had room to operate because he was getting results.[14]

Malcolm's tactics helped turn the Piney Point debate into a win for both the estuary and the organization. While discomfiting to the board, Malcolm's methods made the Piney Point oil refinery an issue at the right time. Citizens in St. Mary's County, along with citizens throughout the region and the country, were beginning to question the policies that supported minimal environmental oversight. Backed by positive press coverage and a united local front, state officials held public hearings, at which residents of St. Mary's uniformly denounced Steuart Petroleum's proposal. Clear public opposition to the plans combined with negative media coverage resulted in the commissioners of St. Mary's County rejecting Steuart Petroleum's proposal. Notably, this was just a few months before the infamous Santa Barbara oil spill. Although Steuart persisted in trying to reverse the decision for another year, the project was dead in the water. In this case, science played a minimal role. The deciding factor was the perceived backroom dealing between Steuart Petroleum representatives and county officials, followed shortly thereafter by a high-profile oil spill. Later, science would occupy a key role in Chesapeake environmentalism, but in its earliest stages the debate had more to do with propriety than with pollution. In this struggle, Malcolm's confrontational rhetoric and demands for accountability and transparency won much praise outside CBF. Fellow Bay advocate Edmund Harvey, president of Delaware Wild Lands, wrote to Malcolm after hearing about the commissioner's decision, "You are to be congratulated for the leading role you played. I consider this decision to be a major victory for those of us throughout the country who are seeking to preserve the marine resources of our countries [sic] estuaries." Harvey added, "I am tremendously pleased that the Chesapeake Bay Foundation decided at this early stage of its existence to take a firm stand in preserving the integrity of the Chesapeake Bay." Harvey was in fact so pleased that he joined CBF's board of trustees and pledged to recruit new members from Delaware. Ordinary citizens were no less impressed by CBF's involvement. A resident of St. Mary's County, Felix Johnson, wrote a thank-you letter to Malcolm in which he said that "of all those who joined with us to frustrate Steuart Oil Company's designs on the County, you gave the most valuable and effective help." Johnson's words carried extra weight because he had helped organize the local opposition to the oil refinery. He added, "I know I speak for all of us who love Saint Mary's when I tell you how very grateful we are." More quantitatively, the cold weight of numbers speaks to the effectiveness of Malcolm's campaign against the Piney Point oil refinery. Starting with only 34 members in May, before the Piney Point campaign began, CBF ended the year with 349 paid-up members, and 1,400 people signed up for CBF's mailing list.[15]

As important as the increased membership and the accolades were, they paled in comparison to the most consequential result of the Piney Point campaign: a $500,000 endowed fund in honor of Tobe Strong's father, L. Corrin Strong. The Strong Memorial Fund allowed CBF to grow into the regional organization its founders had envisioned. Even as late as 1979, a sixth of CBF's budget came from the Strong Fund. Recall that Sherwood's cautiously optimistic goal had been a $200,000 endowment, and his initial request had been denied until CBF could "prove itself." With Piney Point as proof of concept, Sherwood crowed, "No other state or private agency is working as is CBF on a regional basis, entirely for conservation purposes, in the general public's interest apart from the interests of the state, individual developers, commercial users, or corporations." This argument was compelling. The proposal for an oil refinery at Piney Point had been unpopular among the locals but ignored by most Marylanders until Malcolm raised the profile of the issue, demonstrating CBF's value as a regional environmental group that could help champion the causes of smaller, more local groups. This was just the nudge Sherwood and Tobe Strong needed to get the rest of the Strong family to support the idea of an endowment for CBF. The Strongs would continue to be important backers of CBF, including donating property for an environmental education center, but their initial investment in CBF was essential to the organization's growth, stability, and success. More than four decades later, Porter Hopkins reflected that "the funds that they came up with early on was what carried it."[16]

CBF'S RELATIONSHIP TO CHESAPEAKE AND U.S. ENVIRONMENTALISMS

This detailed look at CBF's early years reveals that while CBF was one of several groups that rightly could claim credit for defeating the Piney Point oil refinery, it was the only one to receive a $500,000 endowment for its efforts. Thus, Piney Point illustrates some important fault lines within Chesapeake environmentalism and within U.S. environmental movements. CBF benefitted from being founded by well-connected urban elites, but this is not the whole story. The organization was genuinely innovative, because few people were thinking about the Bay as an ecological unit in the 1960s. Indeed, scientists were only beginning to understand some of the basic mechanisms through which water flowed through the Bay. In other words, if Sherwood deserves credit for leveraging his social connections into a substantial endowment, Malcolm deserves credit for seizing an opportunity to leverage CBF's claims to represent the entire Bay. Although neither man could foresee how far the organization would go, the combination of connections and broad vision would prove powerful and would ultimately allow CBF to be the lynchpin for a movement spanning the six-state, 64,000-square-mile watershed.

In immediate terms, CBF's role in 1968 was to allow the fight over the Piney Point oil refinery to be not just a question of "not in my backyard" but one of "not in my Bay." Citizens in St. Mary's County formed a group, the Potomac River Association (PRA) of St. Mary's Country, but that group never came close to CBF in size and scope. That was partly because there were no Tobe Strongs on the board of the PRA but also partly because no one intended the PRA to be anything more than a voice for locals in St. Mary's County. Seen from the perspective of the residents of St. Mary's, the Piney Point refinery was just another case of NIMBYism. The region needed petroleum, and the location in the Chesapeake was well situated between the important industrial zones of Baltimore and Norfolk. A new refinery had to go somewhere, and there were not many legal tools available to stop it. However, seen from the perspective of the Chesapeake Bay, the refinery threatened not just the quality of life of some people in St. Mary's but the entire public's ability to value the wildlife and recreational opportunities of the Bay. Cynically, one might point out that the region's elites would be most disadvantaged by losing the opportunity for recreational hunting of waterfowl, but then again, it was elite opinion in conjunction with local pressure that was crucial in swaying the St. Mary's County commissioners to reject approval of Steuart Petroleum's proposal.

When scholars look at environmentalism in the United States, they often create a binary between the large national organizations and the small grassroots groups. CBF's regional approach not only allowed it to play a unique role within Chesapeake environmentalism but also made it stand out within the larger ecosystem of U.S. environmental organizations. CBF managed to combine the size and resources of larger organizations with the focus and connection to place of smaller grassroots groups. Again, this too is a reflection of the particular characteristics of CBF's founders and mission. The $500,000 Strong Fund gave CBF the resources to be more than just a grassroots group, but the vision of advocating for the whole Bay region necessitated a group with more capacity and scope than a purely local organization could muster. As sailors, these founders had a unique view of the Bay that was holistic and transcended a particular watershed or tributary, though in fairness, CBF's first years were focused overwhelmingly on issues pertaining to Maryland. Virginia would come later, and Pennsylvania a good while after that. In this and other ways, it was the Bay itself that helped shape a unique collection of environmental groups.[17]

One of the great virtues of CBF's position as a not quite local but not quite national group was that it could intervene effectively in local debates without being labeled an outsider yet still bring significant resources to bear on an issue. Whether it was a power plant on a tributary here or a subdivision draining a wetland there, CBF could effectively transcend local issues and make them relevant regionally, and in some cases nationally. This pattern would play out many times over the next fifty years of environmental activism, but the clearest early example of CBF's dual role was the campaign to stop the Calvert Cliffs nuclear power plant. That particular incident and its consequences for the history of nuclear power in the United States has been explored in detail elsewhere, but a brief overview and a few general remarks will bring some salient points to bear about CBF and its relationship to Chesapeake and U.S. environmental movements.[18]

THE CALVERT CLIFFS CAMPAIGN: THE BAY VERSUS NUCLEAR POWER

The campaign to stop Baltimore Gas & Electric (BGE) from building a nuclear power plant at Calvert Cliffs, Maryland, helped transform the U.S. nuclear regulatory apparatus and created the illusion of a great environmental awakening along the shores of the Chesapeake. It would simultaneously prove beyond a shadow of a doubt that the concept behind CBF was powerful enough to affect national issues while also nearly ripping the organization apart. Jess Malcolm led an increasingly confrontational struggle against BGE along the lines of his aggressive PR blitz against the Piney Point oil refinery with mixed results; the plant was ultimately built, after a thorough environmental review and a landmark court case that tested the National Environmental Policy Act (NEPA) and helped lead to the demise of the Atomic Energy Commission (AEC). As with the Piney Point campaign, he teamed up with local groups and used CBF's position as not quite local but connected to the Bay to make the nuclear power plant a Bay-wide and ultimately a national issue instead of a NIMBY concern about the location of a power plant in a rural countryside (which it also was). However, Malcolm misread the reaction to BGE's plans for a nuclear plant as a broader reaction in defense of the Chesapeake and pushed an increasingly antidevelopment line that alienated him from Sherwood and others on CBF's board, leading to a dramatic power struggle and Malcolm's ouster. In the end, CBF remained an organization ahead of its time, with a broader public concern about the Bay still years away.

While the drama over the Piney Point oil refinery played out in public, CBF's opposition to the Calvert Cliffs nuclear power plant was simmering in the background. Initially, both Malcolm and Sherwood were united against BGE's power plant proposal. For Malcolm, the ecological question of how thermal pollution from the plant would affect the Bay made the decision to build the plant a risky gamble with the Bay's health. Although Malcolm was deeply skeptical of nuclear power, he initially framed his public opposition to BGE's plans around the threat thermal pollution posed to the estuary's ecosystem and the civic argument that "the public has a right to know—*now*—whether or not plants of this type will affect the Bay adversely." In his public comments, Malcolm also hinted at his hard-line environmentalism, noting that "neither I nor anyone else I know, would want cheaper electricity at the expense of any portion of Chesapeake Bay." Sherwood, by contrast, did not play a large public role in the opposition to BGE's nuclear plant, but backed Malcolm from his position on the board. Sherwood was not as ideologically opposed to nuclear power as Malcolm, and he initially tried to form a back-channel relationship with BGE. In his mind, CBF could serve as an "objective conduit of information" that would "keep the public informed" about the nuclear plant. Sherwood eventually secured a meeting with BGE's president and vice president, Edward Utermohl and John Gore, in hopes of establishing a working relationship with the company. The meeting was a disaster. Sherwood tried to persuade Utermohl and Gore that CBF could help the company communicate accurate information about the costs and benefits of nuclear power because CBF would "be less subject to misrepresentation and accusations of self-interest and ulterior motives." Further, Sherwood asked for no fee for this "service"—he saw it as a win-win situation: BGE would get a partner with environmental credentials, and CBF would further raise its profile as the leading voice for the Chesapeake at a time when it desperately needed publicity. It is unclear what Sherwood thought his chances of success were before the meeting, but he was shocked by Utermohl and Gore's response. "Quite incredibly," Sherwood wrote, "I found myself listening to a counterproposal, the substance of which was that if CBF *really* had the public's interest at heart, it would be doing its utmost to support and disseminate official G. & E. statements." Taken aback by how curtly Utermohl and Gore dismissed his offer, Sherwood reported to the CBF board that "it is obvious that there is a need for an objective investigation for the best interest of the public." For the time being, for personal and ecological reasons, Sherwood and Malcolm would present a united front within CBF against the Calvert Cliffs nuclear power plant.[19]

Malcolm would play a pivotal role in the looming Calvert Cliffs campaign because he, and by extension CBF, was the lynchpin in a coalition—the Calvert Cliffs Coordinating Committee, dubbed Quad-C for short—that ultimately sued the AEC and eventually won a landmark court case, *Calvert Cliffs Coordinating Committee v. U.S. Atomic Energy Commission,* which set a broad precedent for the applicability of NEPA and contributed to the dissolution of the AEC. In both these instances, CBF would fulfill its founders' vision of a group that could speak for the Chesapeake region and unite the many smaller organizations already working on the Bay. However, Malcolm's tactics in achieving these victories soon alienated Sherwood, and after a falling out, Sherwood rallied board support to boot Malcolm from the organization. Finally, while Malcolm was able to rouse a potent opposition to the specific threat of a nuclear power plant, the wider public was not ready to embrace the general cause of saving the Bay. Fighting off a scary, science-fiction threat in the form of nuclear power was one thing, but undergoing a large-scale effort to reform industrial society's relationship with the nation's largest estuary was not a step that many, even within CBF, were willing to take.

Before detailing the specifics of CBF's role in the political and legal fight over the power plant, it is worth noting the legitimate concerns and objections to the plant. Like other environmentalists in the late 1960s, CBF was concerned about thermal pollution. Crucially, the environment of the Chesapeake posed several unique challenges to building a nuclear plant. From a technical perspective, a nuclear power plant built along the shores of the Chesapeake could not use the type of large cooling towers made iconic by *The Simpsons* because of the salt content of the estuary's waters. From a cost perspective, buying land for cooling ponds, even in rural Calvert County, would be prohibitively expensive. The elegant solution, from an engineer's perspective, was to simply cycle water from the Bay through the power plant and back out again. Known as a once-through cooling system, this process shielded Bay water from coming into contact with radioactive material through a separate pipe system, reduced the physical footprint of the plant, and was cost-effective. The only pollution would be water slightly warmer than when it entered. However, there were several downsides to this approach. One problem with

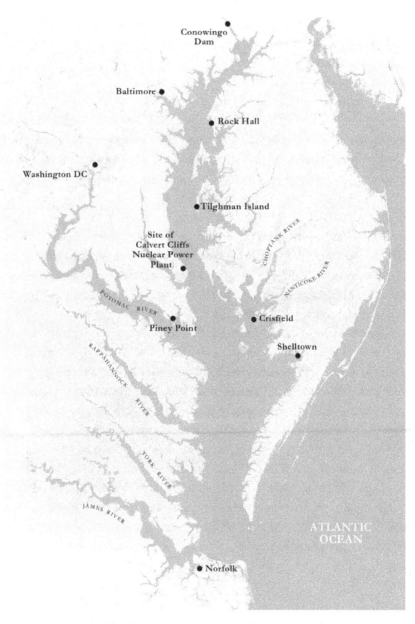

Key locations in the Chesapeake Bay. (Illustration by Turner Blashford, from USGS base map)

this plan was that aquatic organisms are extremely sensitive to temperature fluctuations, and even a fluctuation of a few degrees can cause significant fish kills. The sheer scale of BGE's proposal was also concerning, as it called for the plant to discharge 3.5 billion gallons of water daily into the estuary. Most alarming of all, the physical location of the plant would be at one of the most vulnerable points in the Bay (see map on p. 34). According to the leading blue crab biologist (and a CBF trustee), L. Eugene Cronin, the plant's location at Calvert Cliffs was "the most constricted bottleneck of the Chesapeake, so that crabs and all of the fish species which are obligated to migrate between low salinity and high salinity areas must successfully pass this site, usually at least twice during their life history." The fear motivating opposition to the power plant was that the volume and location of heated discharge could potentially disrupt the entire Chesapeake ecosystem.[20]

This concern about a systemwide crisis was in stark contrast to the concerns of one of the other major players in the Calvert Cliffs campaign, the Chesapeake Environmental Protection Association (CEPA). Unlike CBF, which aspired to a regional role, from its inception in February 1969 CEPA was local and reactionary. In an archetypical example of NIMBYism, CEPA stated as its founding goals "to seek less objectionable placing of power lines which are now proposed for the entire length of the county" and "to seek the relocation, or less harmful effects of a nuclear power plant now proposed to be built on the shores of Calvert County." CEPA was a usefully ally in that it could rally local opposition, but its particular objections to the power plant were not going to go anywhere. At times though, CEPA's concerns muddied the debate about the power plant. A special report on the Calvert Cliffs controversy published in April 1969 by the Annapolis *Evening Capital* claimed that "nothing really got underway until it became known that high-voltage lines from the plant would cut a swath through some scenic and historic Anne Arundel countryside." While the report mentioned thermal pollution, its main focus was on the threat of power lines that would "mar the countryside" and "the terrifying effects a disaster at one of these plants would have." Both of these complaints would be easy for BGE to parry, and neither the threat to local aesthetics nor the possibility of nuclear catastrophe motivated Malcolm and CBF's opposition to the power plant.[21]

Despite increased scrutiny of the plant, its opponents suffered a major setback on June 30, 1969, when the AEC granted BGE a provisional permit to begin constructing the plant's nuclear reactors. To Malcolm and other opponents, the whole process felt like an elaborate rubber stamp. The AEC made its decision on the basis of its finding that "there is reasonable assurance that the proposed facilities can be constructed and operated at the proposed site without undue risk to the health and safety of the public." The AEC did not consider any potential environmental damage the plant might cause because it was not part of its mission to oversee environmental protection. As J. Samuel Walker explained in his authoritative history of the Atomic Energy Commission, "The AEC did not address the questions about thermal pollution because at that point it still considered them to be outside its statutory jurisdiction." Consequently, in the AEC's eyes, the "major contested issue involve[d] the discharge of radioactive materials from the plant, particularly the liquid waste discharge containing tritium," not the effects of billions of gallons of heated water on aquatic life in the Bay. The AEC was referring to the water contained within the nuclear reactor, which periodically had to be replaced; however, this "major contested issue" was only secondary to the arguments against the plant's discharge of thermal pollution. The AEC was only legally obliged to consider health and safety. Thus, even if Malcolm and CBF could have definitively proven to the AEC that thermal pollution from the Calvert Cliffs nuclear power plant would cause irreparable harm to the Bay's ecology, BGE would still have gotten its permit—so long as the plant did not directly threaten "the health and safety of the public."[22]

The Atomic Energy Commission's decision was a setback for CBF, but the AEC hearings had a silver lining: they led to the coordination of increased opposition to the plant, which further boosted CBF's stature in the region. Malcolm reported to CBF's board that the hearings had become a "focal point" of activity. Two days before the hearings were set to begin, CBF hosted a workshop attended by roughly thirty people to prepare for them. Following that workshop, Malcolm continued to meet with scientists and environmentalists to share information and to informally coordinate their plans in advance of the hearing. Afterwards, Malcolm worked in conjunction with Sherwood to continue and expand CBF's 1969 media campaign throughout the summer to include TV and radio appearances, a brief documentary on the Bay, and numerous public speaking engagements, as well as to arrange private meetings with other environmental groups and government officials. Many of these activities resulted from the connections Malcolm made in preparation for the AEC hearings. As he corresponded with concerned citizens across the country, Malcolm realized that the Calvert Cliffs campaign had the potential to become a national issue. On August 11 he wrote to an engineer involved in the antinuclear movement in Madison, Wisconsin, that "it has occurred to me that this particular plant might offer the best opportunity to stimulate a national review of the Nation's nuclear power plant policy." Malcolm's next step, on September 25, 1969, was to invite about sixty people, including "research scientists, political leaders, heads of major conservation organizations and other influential group leaders and concerned citizens," to "formulate a unified approach to the Calvert Cliffs nuclear power plant controversy." CBF funded the initial meeting and planning with the goal of building a large framework so that as many of the interested parties as possible could participate in the campaign without duplicating efforts or sending mixed messages. The strategy paid off as Malcolm met with E. Churchill Murray of CEPA, Ruth Mathes of the Baltimore chapter of the Sierra Club, and a few others on November 19 "to enable an ad hoc committee to work out the organizational procedures for the new group." Those efforts culminated in the formation of the Calvert Cliffs Coordinating Committee on December 10, 1969, with the official election of an executive committee and creation of eight standing committees to set the agenda for the collective effort.[23]

Events accelerated in 1970, as Malcolm and his allies, calling themselves "Quad-C" for short, attempted to halt the construction of the plant through hearings at the state and federal levels. What ultimately turned the Calvert Cliffs campaign into a national issue was the recently passed National Environmental Policy Act, which required that the federal government evaluate every action "which significantly affect[s] the quality of the human environment." Since its creation in 1947, the AEC had only been charged with ensuring that civilian nuclear reactors posed no threat to public health, safety, and domestic security. The AEC insisted that NEPA did not apply to its unique mandate regarding nuclear power, but Malcolm and his allies saw an opportunity to test the new national environmental law. In a March 24, 1970, letter to a fellow environmentalist who was engaged in the Cayuga Lake nuclear power plant controversy, Malcolm revealed that it was Quad-C's intention "to continue pressing the issues that have arisen at Calvert Cliffs for as long and to the extent necessary to make this a test case." The National Wildlife Federation agreed, stating that the case "is a classic example of a national issue. With no legal restraints on siting the Baltimore Gas and Electric Company unilaterally chose a site, received an AEC construction permit based solely on health, safety and national security criteria and plunged into construction." The NWF went on to say that BGE and other utilities planning to build or building nuclear power plants could disregard environmental concerns because of "the Atomic Energy Commission's consistent claim of immunity from considering the environmental impact of activities under its jurisdiction."[24]

With the support of the Sierra Club and the National Wildlife Federation, Malcolm and his local allies in Quad-C petitioned the Atomic Energy Commission on June 29, 1970, for a full review of the Calvert Cliffs plant under NEPA. They had four specific requests: that the AEC order BGE to submit an environmental assessment; conduct its own in-house environmental studies; order BGE to show just cause why construction should not be suspended until the aforementioned assessments were completed; and most critically, "promulgate rules and regulations for applying NEPA to all nuclear power plants for which construction permits . . . have been issued and which have not yet received operating licenses." With the fourth request, Quad-C dramatically changed the terms of the conflict. No longer was this a case of NIMBYism or a regional dispute about a power plant; it was a test case to determine the fate of environmental protection and nuclear regulation in the United States.[25]

Both sides realized that the Calvert Cliffs campaign was in its final stages and that its outcome would set the standard for future action regarding nuclear power. Malcolm called the petition a "turning point" in the battle over nuclear regulation, one whose "importance cannot be underestimated." If the AEC did not conduct an environmental review or force BGE to submit an environmental assessment, then prospects for forcing other utilities to do the same would be bleak. Conversely, if the AEC acceded to Quad-C's demands, it would have to acknowledge NEPA's expansion of the AEC's mission to include environmental protection and review all nuclear plants. Facing such a critical juncture, the AEC did what bureaucracies do best: it delayed, then it stalled, and then in a dramatic move it postponed making a decision. Irritated by the AEC's recalcitrance, Quad-C issued an ultimatum on November 12, 1970, giving the AEC ten days to make a decision or face Quad-C in court. The AEC, maintaining that it was still "considering new regulations" under NEPA, refused to act. Finally, on November 25, 1970, Quad-C, with the Sierra Club and the National Wildlife Federation as co-petitioners, filed a petition for review with the U.S. Court of Appeals for the DC Circuit. They charged that "the continued failure of the Atomic Energy Commission to take action on the original requests of petitioners was tantamount to a denial of those requests." Furthermore, the AEC's intransigence meant that "the public was being locked into the Calvert Cliffs plant as presently planned without any hope of feasible and environmentally more desirable alternatives being accepted." This petition became *Calvert Cliffs Coordinating Committee, Inc. v. AEC*. The final showdown between the two combatants occurred on April 16, 1971, when they argued their case before DC circuit judge J. Skelly Wright.[26]

Quad-C's position was straightforward: BGE and AEC were in violation of the law, and according to NEPA, they had to complete an environmental impact statement before proceeding with construction of the nuclear plant. The AEC argued that although NEPA required environmental impact statements, if the commission had to weigh environmental concerns along with safety and security concerns, it would result in expensive and lengthy delays in licensing atomic energy plants. Moreover, it was the AEC's official position that nuclear power plants were environmentally benign, unlike coal and natural gas plants, which released many types of harmful pollutants into the environment. To the AEC, "meeting the demand for electricity was a more important and immediate problem" than completing the environmental impact statements required by federal law. Judge Wright vehemently disagreed with the AEC's position. In an oft-quoted passage rebuking the AEC, Wright wrote that "the Commission's crabbed interpretation of NEPA makes a mockery of the Act." Furthermore, Wright ruled that the AEC must "take the initiative of considering environmental values at every distinctive and comprehensive stage of the process." Wright's decision was an impressive victory for environmentalists. It forced the AEC to delay issuing new licenses and permits for seventeen months while it made fundamental changes in its atomic licensing process by balancing the environmental effects of a power plant with its safety, security, and economic feasibility.[27]

The *Calvert Cliffs* decision fundamentally reshaped the AEC's mission and also reinforced NEPA's applicability to all federal agencies. Wright's decision has been called "one of the most inclusive, holistic court readings of NEPA" and "a wake-up call to federal agencies that were not routinely implementing NEPA." In explaining his decision, Judge Wright set an important precedent. He said that on the basis of NEPA's legislative history, Congress intended to make "environmental protection part of the mandate of every federal agency." Wright's statement that "environmental protection is as much a part of their responsibility as is protection and promotion of the industries they regulate" reverberated throughout the federal government. The effects of Wright's decision in the *Calvert Cliffs* case were immediate and far-reaching. CBF and its Quad-C allies obtained their environmental impact statement after four years of petitions and protests; while this delayed the plant's construction so that it came online in 1975 instead of in 1973 as originally planned, the most dramatic effects of the case were on the AEC. Following the decision, the AEC added experts trained in environmental fields to its regulatory staff and licensing boards. The AEC's new obligations also resulted in a de facto seventeen-month moratorium, as no plant received an operating license or construction permit while the commission completed its environmental assessments. This period also marked the end of the AEC, as the *Calvert Cliffs* case and other controversies led Congress to abolish the AEC and replace it with a new independent body, the Nuclear Regulatory Commission. Thus, with one dramatic decision stemming from the shores of the Chesapeake, the *Calvert Cliffs* case brought the entire nuclear industry to a screeching halt for nearly a year and a half, contributed to a major regulatory overhaul, and helped to ensure that environmental protection was a factor in all aspects of federal agency decision-making.[28]

CALVERT CLIFFS AND THE FALSE DAWN OF CHESAPEAKE ENVIRONMENTALISM

Tragically, by the time of Quad-C's victory in court, Malcolm had been forced out of the organization he worked so hard to build. Malcolm had misread the situation within CBF and the wider support for protecting the Chesapeake. While the Calvert Cliffs campaign was an early sign that CBF's approach worked and that the regional environmental group could affect both local and national environmental policy, it was also a false dawn for environmentalism in the region. People along the shores of the Chesapeake were spooked by the mysterious new threat posed by nuclear power, even if they didn't quite grasp that the main issue with the plant had nothing to do with nuclear radiation itself. Most citizens of the region, and certainly the majority of CBF's board, were not ready for a broad program of regulating human impacts on the Chesapeake, nor did many yet think it was necessary. Indeed, by the time of Judge Wright's ruling, Malcolm was so disgusted at the way he had been forced out of the Chesapeake Bay Foundation that he retired from public life for good.

Malcolm's problems began, ironically, with the formation of Quad-C. This was something that both fulfilled a long-sought vision of CBF's founders and deeply unsettled them. Back in 1966, the founders had articulated their vision for CBF to be an organization that would have "a major and growing role to play in the future of the Chesapeake Bay Country, as a coordinating group around which the many other civic and anti-pollution and improvement organizations can rally." With the creation of Quad-C, Malcolm had achieved precisely that, and he had helped reshape national environmental and nuclear policy to boot. However, board members like Pritchard, who were put off by Malcolm's aggressive tactics in the Piney Point campaign, were further dismayed by his public feuding with BGE and AEC officials. Publicly, Malcolm blasted BGE officials and even got into a spat with the head of the AEC, Glenn Seaborg, after calling nuclear power "the ultimate form of pollution." By contrast, Sherwood felt that "we *may* fight the Gas & Electric Company over their Calvert Cliffs Project, but we want the company as a member and believe they should be one." Malcolm was fundamentally more willing to be confrontational than Sherwood, and that confrontational nature unnerved CBF's board both because it was not in keeping with the gentlemanly spirit in which they had founded CBF and because Malcolm's antipollution crusade threatened to alienate the region's business community. More than forty years later, Minor Lee Marston, who had been a board member at the time, recalled that Malcolm's opposition to nuclear power had created an unresolvable tension because Malcolm refused to accept nuclear power, whereas most others at CBF "realized that you have to live and you may not like BGE, but you have to live with them."[29]

These internal tensions over tactics and approach boiled over in 1970 as Malcolm ratcheted up the pressure on BGE and began threatening legal action through Quad-C. Malcolm truly believed the revolution had come. On the eve of the first Earth Day in 1970, Malcolm wrote in a special message that "now obviously is the time of reawakening, for a reassessment of values." He often used the quasi-religious term *great awakening* to describe what he thought was occurring in the region and the country. While Earth Day was indeed a profound awakening for many people, the genius of Earth Day (as argued by the historian Adam Rome) was that it was a decentralized, heterogeneous, and uniquely local moment that added up to a national demonstration. On the Chesapeake, this meant that concern for the environment did not manifest as a general concern about the harmful relationship between an industrialized society and the nation's largest estuary. Most people simply did not see the Bay as being critically threatened by society at large. Specific threats, such as oil refineries or power plants, registered as dangerous threats to be stopped, but the broader reevaluation of society's role toward nature that Malcolm sought would have to wait. Indeed, by the end of 1970 Malcolm would ruefully write in his final annual director's report, "I believe that now awakened, we need never again concern ourselves with an apathetic public—save possibly in the area of financial support." As Malcolm's remark implies, the spark lit by Earth Day failed to fuel donations to CBF's cause. Despite the backbone of the Strong Foundation, from the perspective of CBF's trustees, and Arthur Sherwood in particular, CBF's campaign against BGE was proving to be too costly for the organization, in terms of both its finances and its reputation.[30]

On December 12, 1970, with Quad-C's lawsuit against the AEC pending, Sherwood decided to launch his coup to oust Malcolm. His reasons are revealing. In a confidential memo to CBF's board of trustees, he wrote that Malcolm "is not, and does not seem to me to be becoming, the kind of person who is good at sharing authority, working with this Board and many others, getting a good press for Bay and CBF affairs, putting out quality CBF publications, acting as an unofficial ombudsman, gaining custodial responsibility for strategic Bay properties, educating the public—in short, to do the job CBF set out to do." Although Sherwood was upset with Malcolm and Malcolm had aggravated several other board members, it was hard to ignore the fact that Malcolm had achieved a great deal in three years as CBF's director. Where Sherwood could land blows against Malcolm was in how his approach undermined the organization's finances. In his memo, Sherwood continued by saying that Malcolm "has made it difficult for me and this Board to operate effectively in the area of fund-raising. Apparently he has what he himself styles as an 'authority hangup' . . . for whatever causes, I have simply not been able to chart a sensible financial course with him." Indeed, by the end of 1970 CBF was facing a small financial crisis. After Earth Day seemed to herald Malcolm's expected "awakening," he hired two full-time employees in June to help him manage CBF's growing operations. However, sustained revenue streams failed to materialize, and after facing a seven-thousand-dollar gap between revenue and expenses, Malcolm had to fire his first and only two employees, precipitating a showdown with Sherwood over the direction of the foundation.[31]

While Sherwood blamed Malcolm for spending too much money, Malcolm privately complained that knowledge of fundraising "eluded" Sherwood and CBF's board. Malcolm had some legitimate grounds for complaint; for instance, Sherwood himself groused that his fellow trustees "have made it clear that fundraising is simply not their cup of tea." However, for all of his frontline activist work, Malcolm had not built a financial base of support. Despite CBF's well-publicized campaigns on behalf of the Bay, by the end of 1970 it had only nine hundred dues-paying members. The grassroots support Malcolm expected after Earth Day never materialized. Chesapeake environmentalism had not yet coalesced as a force in the region. For as much as Malcolm labored tirelessly on behalf of the Bay, and as gifted as he was at building coalitions, first to contest the Piney Point oil refinery and then to contest the Calvert Cliffs nuclear plant, he never developed a fundraising strategy, nor did he devote much time to recruiting members. Malcolm just expected them to show up in appreciation and support of CBF's good works. When they failed to show up, Malcolm was devastated. In public remarks after announcing his resignation, Malcolm said that "it's very disconcerting to realize that the public isn't willing to support a strong voice in their behalf to protect the Chesapeake Bay from wanton abuse." He added, pointedly, that "if there is a concerned public, it has to show itself. But I've begun to wonder whether 'the concerned public' really exists." Malcolm continued, "It seems incredible that after three years of existence the foundation has only 900 members . . . the number should be at least 90,000." He never quite grasped the difficulty of recruiting and retaining members; for all of its successes, CBF would not exceed ninety thousand members until 2001.[32]

The upshot of all this was that Arthur Sherwood, primarily through his friend Tobe Strong, controlled CBF's financial spigot. Whether Sherwood was truly afraid that Malcolm was driving the organization to financial ruin or just fearful of losing control of the organization is unclear, though some combination of the two is most likely. What is certain is that he took advantage of the financial crisis at the end of 1970 to call for Malcolm's ouster and place himself firmly in control of CBF. Behind the scenes, Sherwood picked his allies carefully. In his memo to the board, Sherwood added a handwritten postscript to Donald Pritchard, who previously had expressed concerns about Malcolm. "Don," Sherwood wrote, "I think its time we stop compounding our errors—I think it was an error to try to make Jess [Malcolm] an administrator." To the entire board, Sherwood wrote in advance of their year-end meeting that "the evidence is clear that we are not fund-raising effectively. To remedy the situation, I'm proposing that we re-structure the administrative staff of CBF, making me the full-time Director." In January 1971 CBF's trustees asked for Malcolm's resignation and voted to name Arthur Sherwood CBF's new executive director. Completing Sherwood's takeover, his good friend and CBF's primary benefactor, Tobe Strong, took Sherwood's place as president of the board.[33]

At the end of the day, Sherwood's connection to the Strong family and their finances meant that he had the winning hand. For all the publicity and indeed meaningful achievements, Malcolm had not been able to tap into a source of funding for CBF that would give him an independent power base in the organization. While he bears some of the blame for this, there is some truth to his observation about public interest in the Bay. Despite some acute threats to the Chesapeake and the national environmental awakening spurred by Earth Day, there was not yet a Chesapeake environmental movement, even if there was a Chesapeake Bay Foundation. Although he retired from public life bitter and disappointed, Malcolm had helped bring CBF through its first few rocky years and given the Strong family reason to agree to Sherwood's pleas for financial support. When the environmental awakening did finally arrive on the shores of the Chesapeake, CBF was in a position to lead the regional environmental movement in large part because of Jess Malcolm's hard work.

CONCLUSION

CBF's ability to affect local and national politics during its early years points to an important conclusion about the U.S. environmental movement, namely, that there was not just one U.S. environmental movement. To understand the policy successes and failures across the country, it is crucial not to overgeneralize about "U.S. environmentalism." It is more accurate to refer to plural U.S. environmental movements, or U.S. environmentalisms, in acknowledgment of the tremendous regional variation in environmental advocacy. The types of environmental movements in the United States are as varied as the nation's ecosystems. The Chesapeake environmental movement could be further subdivided, and indeed, later fissures in the movement would speak to these divisions. However, there did emerge a coherent, uniquely identifiable Chesapeake environmental movement defined in part by the particular characteristics of the Bay as an estuarine environment and in part by the particular characteristics of the society along its shores. This movement to restore the Bay—Chesapeake environmentalism—would repeatedly interact with national environmentalisms, but the movement itself has remained steadfastly regional in nature.

The campaign to stop the Calvert Cliffs nuclear plant illustrates this dynamic brilliantly. At the local level, the Chesapeake Environmental Protection Association fussed about power lines in a classic formulation of NIMBYism. At the national level, organizations across the country raised the alarm about nuclear power and advocated for new federal policies, such as the National Environmental Policy Act. Although lightly touched upon in this narrative, international actors played a role as well, especially in highlighting unanswered questions about nuclear power. Yet what stitched these forces together to form an effective coalition was the regional actor, the Chesapeake Bay Foundation. Regionalism in the U.S. environmental movements has too often been overlooked, and as this analysis has shown, regional actors play a critical role. They are big enough to evade charges of NIMBYism but sufficiently connected to local communities to avoid charges of being troublemaking outsiders. In this way, CBF would become an engine of environmental advocacy for the next half century.

Unfortunately for Jess Malcolm, the Chesapeake restoration movement was still in an embryonic phase, lagging behind the environmental awakenings occurring in other parts of the country. Years later, the shadows of the Malcolm era lingered in Sherwood's mind. In CBF's 1976 annual report, he wrote "If there is one thing we fear almost as much as irreparable damage to the Chesapeake estuarine system, it is identification with the emotional hothead who damns everything human while pursuing his own vices." The challenge of sounding a klaxon about the dangers facing the Bay without appearing to be irrationally opposed to industrial, agricultural, and commercial uses of the Bay would be one of the defining features of Chesapeake environmentalism. As much as anything else, what set the Chesapeake apart from other ecosystems in the twentieth century was that the movement's goals were never about protecting a "pristine" wilderness from the stain of human touch. The overriding challenge of the Chesapeake, which makes it so compelling and so worthy of study, is balancing the needs of the ecosystem with the needs of the people that depend on it for work and play. When Malcolm left the foundation in January 1971, he was convinced that the Bay's relationship with its human population was dangerously out of balance. He was right. The following year, Tropical Storm Agnes would devastate the region and dramatically vindicate Malcolm's concerns. What Malcolm could not have foreseen when he left CBF defeated and dejected was that the people of the region would, in fits and starts, rally to the cause of saving the Chesapeake and, against the odds, succeed.[34]

TWO

THE STORM

Historians are fond of periodization, and with good reason. Breaking the seamless flow of history into discrete chunks is essential to crafting a compelling narrative with a beginning, middle, and end. One famous example of periodization—to historians, at least—is the so-called long nineteenth century. While a nonhistorian might quite rightly assume that the nineteenth century began in 1801 and ended in 1900, most historians choose to extend the century (centuries themselves being basic forms of periodization) to cover the events of the French Revolution as well as World War I. By bookending the century with these two cataclysmic upheavals, historians can better illuminate the European system created by the first and destroyed by the second. Likewise, this chapter argues for the need to extend the conventional narrative of the Chesapeake Bay's environmental movement in order to present a broader, more accurate, and ultimately more informative history of how the movement began, who the important players were, and why it created the groundwork for saving the Chesapeake.

The conventional narrative of Chesapeake environmentalism contains several major flaws. Most problematically, it excludes the voices of important historical actors and marginalizes the role of the Bay's ecological health as a driving force for the movement. The dominant narrative dates the beginning of Chesapeake environmentalism to the early 1980s, in particular to 1983, when the first Chesapeake Bay Agreement was signed and the Chesapeake Bay Program was formalized. The momentous events that immediately preceded the signing of the first Chesapeake Bay Agreement are explored in depth in the next chapter, but they must first be contextualized as a critical *turning point* in the movement, not its beginning.

As we have seen, the Chesapeake Bay Foundation was active from the mid-1960s, while the broader movement was still in an embryonic stage. While that history is relevant and valuable, as Jess Malcom learned the hard way, Chesapeake environmentalism had not yet emerged by the time of the first Earth Day in 1970. The beginning of Chesapeake environmentalism might better be located in the aftermath of Tropical Storm Agnes, which struck the region in 1972 and caused widespread economic and environmental damage, from the latter of which the Bay still has not fully recovered. The record-setting rainfall that Agnes dumped across the Bay's watershed precipitated a sudden and dramatic decline that caused a scientific and political reaction that culminated a decade later in the first Chesapeake Bay Agreement. To better understand this history, we will look through the eyes of both small-scale commercial fishermen, or watermen, and scientists studying the estuary to understand how Agnes profoundly altered their understanding of the Bay and how that gave rise to Chesapeake environmentalism. Thus, the Chesapeake's watershed moment was not when the region's political leaders agreed to take the first tentative steps toward a coordinated approach to the Bay's recovery in 1983, but rather a decade earlier, when Tropical Storm Agnes laid bare the Bay's fragility.

THE CONVENTIONAL NARRATIVE: SENATOR CHARLES MATHIAS AND THE ENVIRONMENTAL PROTECTION AGENCY

Before launching into a new periodization of Chesapeake environmental history, it is important to first understand the old one, especially its shortcomings. The dominant narrative of the Chesapeake's restoration movement goes roughly as follows: Senator Charles Mathias (R-MD), after hearing the

The conventional narrative of Chesapeake environmentalism, represented by Senator Charles Mathias receiving an award in 1983. *From left,* Senator Mathias; Cranston Morgan, chairman of the Citizens Program for the Chesapeake Bay (later renamed Alliance for the Chesapeake Bay); and Godfrey Rockefeller, president of the CBF board. (Courtesy of Chesapeake Bay Program)

complaints of his constituents, successfully pulled the levers of power in Congress to secure funding for a major multiyear EPA research program that would determine once and for all what was wrong with the Bay and what could be done about it. The results of this research program, informally known as the Mathias study, painted a stark picture of the Bay's health. Concerned citizens demanded that their elected leaders take action, so in 1983 officials from the key watershed states convened for a conference on the Bay. The conference ended with representatives from Maryland, Virginia, Pennsylvania, DC, and the EPA signing the Chesapeake Bay Agreement. For the first time, these three states and the federal government pledged to cooperate "to fully address the extent, complexity, and sources of pollutants entering the Bay."[1]

This conventional narrative appears with minor variations in accounts by scholars, journalists, and activist groups and in government reports. Perhaps no example better illustrates how most sources portray the dominant narrative than the first one that appears when one does a Google search for the history of the Chesapeake Bay's restoration: the Chesapeake Bay Program, the very institution created by the Chesapeake Bay Agreement in 1983. Its website reads: "In the late 1970s, U.S. Senator Charles 'Mac' Mathias (R-Md.) sponsored a Congressionally funded $27 million, five-year study to analyze the Bay's rapid loss of wildlife and aquatic life. The study, which was published in the early 1980s, identified excess nutrient pollution as the main source of the Bay's degradation. These initial research findings led to the formation of the Chesapeake Bay Program as the means to restore the Bay." This is the classic formulation of the dominant narrative, and it is unsurprising that the Bay Program would place its own creation at the center of the narrative. However, this is no mere case of self-promotion. Most other accounts of the Bay's restoration movement follow this timeline.[2]

Tom Horton is widely acknowledged as the Chesapeake's foremost environmental journalist, one of the leading authors of nonfiction works on the Bay, and a prominent voice for environmental protection. In the 2003 update to his book *Turning the Tide* he asked, "What lessons do we draw from the paltry results of an 'acclaimed' restoration attempt that has been operating in earnest for nearly two decades?" Other observers were also questioning the progress of the restoration movement around the same time. Maryland's former governor, Harry Hughes, said in 2002, "Frankly, I would have hoped that it would have been further along than it is after 18 years." Likewise, in a 2003 book the political scientist Howard Ernst wrote a critical analysis of the Bay restoration and came up with a slightly earlier date. "The restoration effort's political structure began to take shape in 1980," he wrote, "when policy makers in Virginia and Maryland came together to create the Chesapeake Bay Commission."[3]

These examples, collectively representing journalism, government, and scholarly sources, have two important aspects in common. First, they all date the beginning of the movement to the early 1980s. Although there is minor variation, the general consensus is that events kicked into gear in the early 1980s. The other, more important aspect is the profound sense of disappointment in the accomplishments of the restoration movement. By the early years of the twenty-first century the Bay region faced both political and ecological setbacks, and the actors who had been struggling for the past two decades wondered what had gone wrong. Indeed, there are echoes of that period in current sources. CBF's timeline for the Bay's restoration also begins in 1983, with the following comment, "Prior to 1983, with rare exceptions, the jurisdictions that make up the Bay watershed made their own plans and programs independent of one another. Even after the first joint agreement was signed in 1983, efforts remained voluntary—and, unfortunately, ineffective." The conventional narrative is so powerful that even one of the institutional actors that preceded the work of the 1980s continues to use 1983 as a critical reference point. A new narrative is sorely needed.[4]

This frustration over the ineffectiveness of the restoration is understandable and not without some justification, but it also reveals one of the drawbacks of beginning the narrative in the early 1980s. Beginning the narrative in 1972 highlights the decade-long struggle to get the region's political structure committed to the Bay's restoration. Some of the frustration can be tempered with an understanding of how difficult it was to get the watershed states to begin collaborating in the first place. A shared celebration of that accomplishment can serve as a balm for current tensions and encourage future collaboration instead of the recriminations and finger-pointing that appear to be the region's default setting. Much has been made of the voluntary nature of the Bay Program, and while those critiques have merit, they miss the critical fact that in order for there to be a regional consensus to restore the Bay, it had to be voluntary. After all, a consensus cannot be mandated; it must be achieved. Furthermore, the current legal mechanisms of enforcing Chesapeake cleanup rely heavily on the legacy of partnerships that emerged from the Bay Program. A broader history that places the Bay Program in its proper historical context can accommodate understandings of it both as a messy political compromise and as a major achievement on the Chesapeake's road to sustainability.

One other important drawback to the conventional narrative is that it excludes important historical actors and narrows our conception of who is part of the Bay restoration movement. At its worst, the dominant narrative can read as a "great man" history of Chesapeake environmentalism, with Senator Charles Mathias portrayed as the prime mover of the restoration movement. As deserving as he is of credit, focusing on Mathias can place too much emphasis on political processes and the decisions of government officials, to the exclusion of many people who knew the Bay intimately—scientists and watermen foremost among them but citizen activists as well—and had first raised the alarm about its decline. The question of who makes history is not just an academic one; it is central to thinking about the future of Chesapeake environmentalism and ultimately the health of the Bay. If the history of the Bay's recovery is depicted as the work of a few brilliant politicians, we may despair of ever seeing meaningful policies and legislation move through the hyperpartisan political system. On the other hand, if the Bay's recovery is understood as a process of collaboration between Bay experts, ordinary citizens, and their elected representatives, then we may discover that we possess the ability to further preserve and protect the Bay.

FALSE STARTS, DEAD ENDS, AND A NEW BEGINNING

One final reason to be skeptical of the dominant narrative is that it makes an assumption about historical causality that is far too simple. The many previous conferences about the Bay had not yielded a sustained environmental movement or a new set of policies, and even the "historic" Mathias EPA study was not the first multimillion-dollar study of the Bay. Both scholars and journalists have pointed with derision to the $15 million boondoggle begun in 1965, the Chesapeake Bay Hydraulic Model, which the Army Corps of Engineers spent a decade building at Matapeake on Maryland's Eastern Shore. The Corps mothballed the project after just three years, making the Bay Hydraulic Model an easy target for critics of federal waste and inaction. Yet those critics missed a much more interesting set of questions: If Congress had already agreed to spend millions on the Bay, why was a new study warranted? Why did the Mathias-sponsored study achieve public support for restoring the Bay, when a decade prior the Corps' work on the Hydraulic Model had yielded little to nothing of value? The dominant narrative has no good answers to these questions. In fact, Mathias's study was only the latest in a long series of attempts to coordinate government action on the Bay, and from the perspective of 1973 (or as we will see in the next chapter, from the perspective of 1981), there would have been no clear reason to assume that Mathias's study would fare better in prompting action by policymakers than previous attempts. To understand what made Mathias's study different, we have to look to the Bay itself and the people who knew it best. Tropical Storm Agnes is central to this explanation because it triggered the chain of events that led to the Chesapeake Bay Agreement, leaving so many other attempts to coordinate Chesapeake environmental policy as mere false starts and dead ends. A brief review of some of these previous efforts will help underscore the significance of Tropical Storm Agnes and allow us to recenter an environmental factor in the Chesapeake's environmental history.[5]

Regional concern about the Chesapeake's health long predated the Mathias study, and indeed, the dominant narrative obscures much about the longer history of regional efforts to protect and manage the Bay. The historian Christine Keiner notes in her superb analysis of the Maryland oyster industry that "scientists have been sounding the alarm for more than a hundred years." The Bay's first environmental laws go back even further, dating from at least the early nineteenth century, when legislators in Maryland outlawed punt guns, infamous, artillery-sized shotguns used by market hunters to bring down scores of migrating waterfowl. The state also famously restricted gear that could be used to harvest oysters, thus ensuring that a sail-powered commercial oyster fleet would survive into the twenty-first century. Over the course of the nineteenth century, concerns about the Bay broadened, from natural resource management and specific conservation issues to questions about the ecosystem as a whole. In 1925, Maryland created the Chesapeake Biological Laboratory (CBL), making it the oldest state-supported marine laboratory on the East Coast. Virginia followed suit fifteen years later when it created the Virginia Institute of Marine Science (VIMS), but even so, cooperation between the institutions and a regional awareness of the Chesapeake's environmental problems was limited in the first decades of the twentieth century.[6]

The first meaningful attempt at a regional approach to the Bay came on October 6, 1933, fifty years before the Chesapeake Bay Agreement was signed. Representatives from Maryland, Virginia, DC, Delaware, and the federal government gathered in Baltimore under the auspices of the New Deal's Works Progress Administration for what was most likely the first regional conference about the Bay. George Radcliffe, the WPA regional adviser and soon to be a U.S. senator from Maryland, opened the conference. His remarks succinctly illustrate the overall state of Chesapeake management in the 1930s and the problems facing coordinated action. Radcliffe began the conference by noting that

> we have been working on this matter of the Chesapeake Bay for a long time, everybody has been interested in it for many years, and lots of us have been doing things in regard to it.... We felt that the time had come to make a survey of just what is what, that the time had come for all of us who were interested, or as many as we could get here who were interested, to come together and talk of Chesapeake Bay as one interest. The people of Virginia have their interests, those in Maryland have their interests, those in Delaware have their interests, and the same is true with respect to Washington [DC] and Pennsylvania. A good many people who are interested in the matter have been following out their own ideas and views to a certain extent, and we believe there is a decided advantage to be gained by all of us getting together and looking at Chesapeake Bay, first, as it is now, taking stock of what has been done, then giving further consideration as to what can be done.[7]

It is remarkable that Radcliffe began by asserting that people had been working on the Chesapeake "for a long time." This comment reinforces the idea that even as early as 1933, concerns about the Chesapeake were old news. What was new, according to Radcliffe, was that the time had finally come to "talk of Chesapeake Bay as one interest." This likely marks the first time that any government official made an effort to address the Bay as a whole, rather than one of its subcomponents, such as oysters, sewage, or waterfowl, among other specific concerns. Most relevant to this history, Radcliffe noted that people from the different watershed states had their own "interests" and that they had been following their own ideas, rather than working collectively to take stock of the Bay and to consider what could be done about it. While the conference goals were noble, the tension between the different "interests" and looking at the Chesapeake as "one interest" would foil efforts to protect the Bay for another fifty years, and they still hinder restoration efforts to this day. The conference also foreshadowed later events in that it was only under the auspices of the federal government that the various watershed states sent representatives for a regional meeting. It would be as accurate in 1983 as it was in 1933 to summarize the state of Bay management as Radcliffe did, noting that people had been concerned about the Bay for a long time, and while there would be a "decided advantage" in getting together, different interests continued to follow their own ideas.

Although the 1933 conference was the first time that some watershed states and the federal government met to try to coordinate their policies regarding the Chesapeake, it was not the start of a further conservation or environmental movement, nor did it lead to any broader coordination over Chesapeake issues. No major legislation or policies emerged as a result of the conference. Crucially, Maryland and Virginia continued to follow separate Bay policies and at times clashed violently over its resources. Incredibly, the last death in the infamous "Oyster Wars" between Maryland and Virginia natural resource police and watermen came in 1959! However, undercurrents pointing toward greater cooperation emerged in the scientific community after World War II. In addition to the Chesapeake Biological Laboratory and the Virginia Institute of Marine Science, a third major scientific institution appeared on the Bay in 1947, the Chesapeake Bay Institute (CBI) at Johns Hopkins University. The CBI came to be directly as a result of advocacy from scientists at the first two institutions, who saw a need for more research on the Chesapeake and leveraged their respective states' mistrust of each other to appropriate funding for a third, neutral institution outside the two states' public institutions. On the eve of the great national environmental awakening in the late 1960s, harvests of finfish and shellfish generally declined, but Maryland and Virginia continued to point fingers at each other instead of making meaningful progress toward collaboration.[8]

A striking example of the divisions between Maryland and Virginia occurred in 1968, when Maryland's Spiro Agnew convened a Governor's Conference on Chesapeake Bay. In many ways this represented a step in the wrong direction, as no prominent Virginia officials attended or presented. It was an approach to preserving the Bay by Maryland and for Maryland. Yet the event was not without prominent dissenters. Rogers C. B. Morton, a congressman from Maryland's Eastern Shore and soon to be chair of the Republican National Committee, was a strong early supporter of a regional approach to the Chesapeake. He gently voiced disapproval of Maryland's go-it-alone policy, arguing that "in the cold light of analysis, I think we can accurately and truthfully say the system we have employed thus far in managing the resource has been inadequate, and must be revised." In particular, that revision needed to be "a system of management which will develop total involvement of all interests represented and all levels of government . . . for the whole basin." Morton acknowledged that to achieve such an ambitious plan, "undoubtedly, it will be necessary to cross historical barriers and reach for new ideas in regional management and regional planning." Unfortunately for Morton and those who shared his views, there was no political will in Maryland or Virginia, much less in Pennsylvania or the federal government, to cross historical barriers to create a regional approach to the Chesapeake. It would take the worst natural disaster ever to hit the region—Tropical Storm Agnes—to spur the parties to begin in earnest to treat the Chesapeake as "one interest."[9]

TROPICAL STORM AGNES

Although we live in an age of natural disasters increasingly driven by climate change, Tropical Storm Agnes still ranks as one of the worst storms in American history, and certainly in the Mid-Atlantic region. As meteorologists and fans of the Weather Channel know, what makes a tropical storm dangerous is not the wind but the water. Agnes's wind speeds never exceeded what an average major league pitcher could reach with his fastball, and by the time it reached the Chesapeake watershed, it barely qualified as a tropical storm. However, according to NOAA, the rainfall from Tropical Storm Agnes unleashed "the greatest flood to ever devastate the Mid-Atlantic in both coverage and magnitude." At the time Agnes was the costliest storm to strike the United States. Its floodwaters caused billions of dollars in damages and more than eighty deaths in the Chesapeake watershed alone. On top of the sheer volume of water, the storm's path ensured maximum damage to the Bay. The heaviest rains fell straight down the middle of the Susquehanna River basin. If Agnes had been a bullet, it would have hit the Chesapeake right between the eyes. Agnes triggered an ecological decline that the social and political structures of the region have been responding to for the past five decades. Although it would take several years for the damage done by Agnes to sink in among the broader population, the harm to the Bay was immediately apparent to the two groups who knew the Bay best: watermen and scientists.[10]

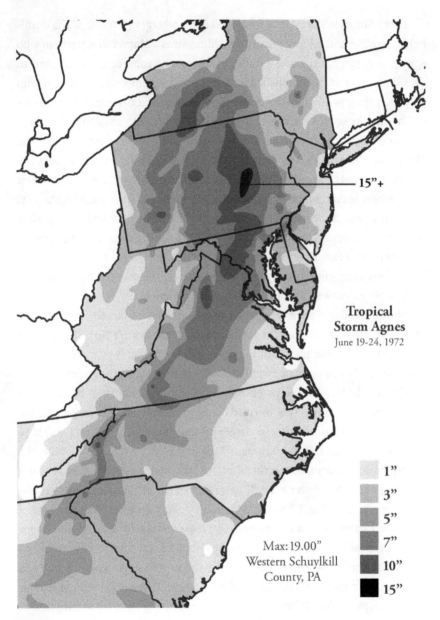

Rainfall from Tropical Storm Agnes. (Illustration by Turner Blashford, from NOAA base map and weather data)

Larry Simns is the type of historical actor who gets left out of the dominant narrative of the Bay's restoration. Simns was a Maryland waterman who spent his entire life making a living by "following the water," as he liked to say. Simns was an accomplished and respected commercial fisherman who became a tireless advocate for watermen and was an on-again, off-again ally of Chesapeake environmentalists. Simns had a remarkable vision and a knack for leadership and organizing. He was a cofounder and the president of the Maryland Watermen's Association (MWA), which he led for forty years before retiring near the end of his life. Simns left a memoir, *The Best of Times on the Chesapeake Bay*, which describes the experience of watermen on the Bay. With the important caveat that this is a memoir told to a coauthor long after the events described, it stands as one of the few firsthand accounts from a waterman who experienced Agnes's impact on the Bay. No account of Chesapeake environmental history would be complete without the perspective of those whose lives and livelihoods depended on the Bay.[11]

Simns recalled a harrowing thirty-six hours riding out the storm in his workboat. He remembered that at the peak of the storm's fury "I could not see anything outside my cabin windows" and "I couldn't even hear my engine idling." Simns was not afraid for himself, for he knew he rode in "a rock solid forty-six foot wooden workboat that had seen plenty of bad weather and angry seas." Simns was worried about his family and his community of Rock Hall, in the northeastern corner of the Bay, which was vulnerable to damage from the storm. After the winds and rain passed, Simns piloted a course home, anxious to see the effects of the historic storm. "On the first morning back," Simns recalled, "I was relieved the physical damage was less than expected. So far, I felt our community could weather the aftermath. I had seen lots of fresh water before, so I wasn't too worried. Up until 1972, the Chesapeake Bay always had the capacity and strength to bounce back." Simns's optimism would not last. The next morning, he woke up early to watch the sun rise over the Chesapeake, grateful that his family and friends had been spared. What he saw in the early morning light made him sick to his stomach. "Right there before my eyes," Simns remembered, "my beloved Chesapeake Bay was turning into a brown, murky mess that can best be described as 'beef gravy.'" Thick brown sediment stirred up by the storm roiled the Bay, and Simns saw everything from plastic bottles to entire barn doors float by. "For a long time," Simns mused, "I couldn't get these images out of my mind." Haunted by what he saw, he "became somewhat fixated on the downstream issues associated with dirt, and gradually I developed some new concerns . . . once I grasped how extensively Agnes assaulted Maryland with excess volumes of fresh water and dirt, I instinctively knew that the plan for my life was about to change."[12]

It is fair to ask with some degree of skepticism whether Simns really knew his life was about to change in 1972, but that would miss the point. The value of Simns's memoir is that after living a life on the Bay intimately connected to the water and spending most of his adult life advocating for watermen, Simns identified Agnes as the Bay's turning point. Whether or not he knew it so clearly at the time matters less than the fact that after reflecting on his career on the water, Simns saw Agnes as the moment when the Bay's fragility was revealed. Simns's account does not speak for all watermen across the Bay, but given his position as the leader of the Maryland Watermen's Association for more than forty years, his point of view carries significant weight.

AGNES AND THE EVOLUTION OF CHESAPEAKE SCIENCE

While Larry Simns was riding out the storm near Rock Hall, at the opposite end of the estuary a group of scientists and citizen activists were gathered in Fredericksburg, Virginia. As is so often the case, their voices are much better preserved in the historical record. The Chesapeake Bay is one of the most studied bodies of water in the world, with sustained academic and governmental research programs stretching back to the mid-nineteenth century. By the time Agnes hit the region, Chesapeake science was dominated by a ruling triumvirate of veteran scientists who headed the region's three main research institutions: Eugene Cronin, of the University of Maryland's Chesapeake Biological Laboratory, William Hargis, of the Virginia Institute of Marine Science, and Donald Pritchard, of the Chesapeake Bay Institute at Johns Hopkins. These three men played a crucial role in the coordination of Chesapeake Bay research, and their informal cooperation laid the groundwork for what would eventually become the Chesapeake Research Consortium. Tropical Storm Agnes would spur them to formalize their coordinated approach to Chesapeake science, which in turn would yield important new insights about why the Bay was in such trouble.

At the time Agnes hit the region, Cronin, Hargis, and Pritchard had little idea of the impact of agricultural nonpoint source nutrient pollution on the health of the Bay. They addressed the subject some years later in a panel discussion organized by the Chesapeake Bay Foundation, which was fortuitously caught on film. When asked what the Bay's scientific community knew about runoff from farms, Cronin admitted that they "did not recognize the importance of agricultural inputs." Hargis concurred, saying, "We were aware that agricultural practices had an impact on water quality, but we didn't know how much." Pritchard defended their record, explaining that they had been aware of the danger of overfertilizing the Bay but believed that the greatest source of nutrients was untreated sewage, not agriculture. Pritchard elaborated, "If you go back and try to explain some of the changes that one has observed, they are related in part to changes in agricultural practices which were occurring concurrently with our looking at the system. You know, Monday morning quarterback you could say 'You should have known that.' But we weren't students of agriculture. And we just assumed, 'well, the farmland is there and it's doing the same thing they've always done.'" Agnes forced scientists who had previously been concerned about the Bay's waters to begin to look at what was happening on land and, if not become students of agriculture, collaborate with scientists who were students of agriculture to ask new questions about how agricultural runoff affected the Bay.[13]

Archival evidence from the time confirms that the recollections of these three scientists were accurate. In 1970 Cronin presented a white paper at a Chesapeake Bay Biological Research Planning Conference (yet another of the false starts of Chesapeake environmentalism). His paper contained an exhaustive analysis of the major problems facing the Bay and its research needs. Cronin compiled his report in consultation with scientists throughout the region, including Pritchard and Hargis, and noticeably absent from his report was any discussion of agricultural nutrient pollution. Cronin listed what the Bay's scientific community had identified as the greatest threats to the Bay's health. This list included thermal pollution from nuclear power plants, construction along the shore and resulting destruction of wetlands, raw sewage, erosion and sedimentation, toxic chemicals, solid wastes, and small craft pollution (i.e., wastes from recreational boaters dumped into the

Bay). This list suggests that as far as Cronin and other Bay scientists were concerned, the major threats to the Bay were from activities in the water and on the shoreline. The broader watershed factored minimally into their understanding of the system. In classic scientific fashion, Cronin called for more research in many areas, including "the cycling of nutrients and other important components [that] are not understood and can be brought to a far higher level of comprehension and use." Chesapeake scientists on the eve of Agnes knew they needed to better understand the Bay's nutrient cycle, but agriculture was not yet on their radar.[14]

Even as rainfall from Agnes was pounding Hargis, Cronin, and Pritchard in Fredericksburg, they were already plotting to take advantage of the opportunity. The three men had previously established an informal mechanism for cooperation that they called the Chesapeake Bay Research Council to collaborate on "data collection, data storage, and data sharing." The three scientists and the institutions they led developed tighter connections while working together to gather data for the U.S. Army Corps of Engineers Chesapeake Bay Hydraulic Model. Cronin later explained the division of labor, telling his audience that "we divided the Bay into three parts, and [VIMS] did the field work, currents, salinity, temperature, and the rest of it for the lower Bay, Chesapeake Biological Laboratory did the center of the Bay, and [CBI] did the upper Bay, all in very neat cooperation, coordination, and interaction." When Tropical Storm Agnes struck, the three men, recognizing the importance of studying the event as it happened, agreed to deploy their collective armada of research vessels and scientists to study the storm and its effects despite lacking any authorization or identified source of funding. In an ironic twist, many of the ships, staff, and equipment used to study Agnes were originally intended to be used for research to support the Bay Hydraulic Model, which quickly became obsolete in part because of what was learned from studying the effects of Agnes. Looking back, Hargis joked that while they never got reimbursed for all their expenses, "we didn't go to jail for it either."[15]

Cronin, Hargis, and Pritchard's efforts to organize a field study of Agnes as the storm hit the region would later be dubbed "an unprecedented scientific community response to an unprecedented event." Their research, which the triumvirate somewhat melodramatically called Operation Agnes, yielded two critical results for Chesapeake science. First, it allowed the ad hoc procedures developed by Cronin, Hargis, and Pritchard to crystallize into a formal organization, the Chesapeake Research Consortium (CRC), which has since helped to facilitate a more collaborative regional approach to Chesapeake science. Along with the Chesapeake Bay Foundation, the CRC was one of the few institutions that took a genuinely regional approach to the Bay at a time when policies and funding for the Bay were still severely Balkanized. The formation of the CRC was in an embryonic phase, spurred by a February 1972 grant from the National Science Foundation when Agnes hit the region. The tropical storm thus significantly expedited the process of collaboration. After getting their teams on the water to start the research, Hargis, Cronin, and Pritchard transferred control to the CRC for completion and publication of findings. All three were heavily involved with the CRC—Pritchard and Cronin were its first two directors—and wanted to see the scientific community study the Bay more holistically. Agnes provided the opportunity to cultivate a more coordinated regional approach to the increasingly complex and costly scientific research on the Chesapeake Bay.[16]

The second critical result from Operation Agnes was the research itself, which yielded alarming new information about the effects of land-use changes in the watershed on nutrient pollution and sediments in the Bay. The major findings of the research were first presented in a symposium in 1974 and then published in a peer-reviewed volume two years later. The preliminary results of Operation Agnes have since been confirmed repeatedly, with increasing detail and nuance, but in short, Operation Agnes showed that contrary to previous assumptions, far more nutrient pollution was coming from agricultural sources than had been anticipated, and far fewer nutrients were being removed from the Bay. According to the final, peer-reviewed publication, "The most striking result of the nutrient fluxes calculated for the Bay mouth is the relatively small extent of nutrient flushing . . . it is apparent that in times of normal streamflow, vastly greater amounts of nutrients are being added to the Bay than are being removed at the Bay mouth." Contrary to the prevailing opinion, the Bay did not "flush" into the Atlantic Ocean nearly as much as had been expected. Furthermore, even during normal streamflow, much less a tropical downpour, nutrients and other chemicals would remain trapped in the Bay, affecting the estuary long after they first washed into its waters.[17]

Tropical Storm Agnes dramatically revealed to both watermen and scientists, in different ways, that the Bay they thought they knew had changed and was far more vulnerable than they had realized. The scientists who dedicated their lives to understanding the Bay came to the same basic conclusion that Larry Simns did, namely, that the Bay wouldn't "bounce back" as it always had. Thus, Tropical Storm Agnes was the turning point for the Bay's health and the beginning of the movement to restore it. Looking back, Donald Pritchard described why he thought Agnes had been so significant for the Chesapeake. He noted that pressures had been building on the Bay's ecosystem and testing its resilience and "then comes along a year like Agnes ... when you get that much of a load upon a slowly changing system, suddenly you trigger—I mean, it's stressed—and you trigger that reaction."[18]

FROM REGIONAL SCIENCE TO REGIONAL POLITICS

As the preceding sections have shown, both watermen and scientists understood Agnes, albeit from different perspectives, as a devastating event with far-reaching potential to reshape the Chesapeake. It is now time to bring Senator Charles Mathias of Maryland back into the story. Charles McCurdy "Mac" Mathias Jr. was born in 1922 in Frederick, a midsized town in rural west-central Maryland. He grew up as a member of a staunchly Republican family that claimed some of the first Republican state legislators in Maryland as part of its lineage. After graduating from Haverford College in 1944, Mathias served during World War II in the Pacific theater as an ensign in the U.S. Navy. After the war, Mathias earned his law degree from the University of Maryland in 1949. Fatefully, Mathias's law school roommate, Daniel Brewster, would go on to represent Maryland in the U.S. Senate before Mathias defeated him in the 1968 election. Senator Mathias built his reputation as a moderate Republican with a strong record on environmental protection and civil rights. Prior to his involvement with the Chesapeake, Mathias had played a pivotal role in helping to preserve the C&O Canal as a National Historical Park. Mathias was deeply respected on both sides of the aisle, with Democratic Senate Majority Leader Mike Mansfield once calling him "the conscience of the Senate." Mathias would eventually fall out of favor with the Republican Party as it came to be increasingly dominated by conservative political ideas. However, in the early and mid-1970s his credibility and influence with both parties were at their peak, allowing him to act as the Chesapeake's champion in Congress.[19]

One key piece of Mathias lore was Mathias's five-day, 450-mile tour of the Chesapeake in response to Tropical Storm Agnes, which supposedly led directly to the EPA study and the Chesapeake Bay Program. This is misleading for several reasons. First, Mathias's tour came a year after Agnes, and although the full scientific results of Operation Agnes were not yet complete, it was becoming increasingly clear that the Bay had not recovered. Second, as we have seen, prior to Mathias's involvement there had been significant federal investment in scientific studies of the Chesapeake, including millions spent on the model at Matapeake, and it had not yielded anything close to a coherent plan to protect the Bay from further damage, let alone restore it. Third, when Mathias began his tour in Baltimore Harbor, he told the press gaggle that his goal was to "determine if new federal legislation is required to protect and improve the bay." The resulting study would in many ways be a consolation prize. Thus, the Mathias tour is better understood as responding to constituents' demands rather than driving events. Indeed, while he began his tour by meeting with executives from Bethlehem Steel, Mathias met with stakeholders from throughout the region, including Larry Simns. Mathias met with Simns on the waterman's workboat, where they talked for hours. Simns remembered Mathias's visit positively and praised his candid style and willingness to listen. Simns recalled, "I told him how the water quality had changed since I had first observed it in the early forties. I also told him all the major issues that fellow watermen from across the state had shared with me as the MWA president. Surprisingly, he did not interrupt me or get defensive when I told him less than flattering truths about his (and my) beloved state of Maryland. He did ask me questions to further his understanding but did not try to reverse or change my beliefs." Unsurprisingly given his political success, Mathias had a gift for communication and for winning trust. Simns saw Mathias as an ally throughout their careers, and despite their at times fierce disagreements, when Mathias retired from the U.S. Senate, Simns was among the keynote speakers at a banquet in the senator's honor.[20]

The Mathias tale is also incomplete because it obscures the roles of other actors, even other prominent national figures. We have already met one important and underappreciated player in this story, Rogers C. B. Morton, a prominent Maryland Republican who had called for a regional approach at the 1968 Maryland Governor's Conference on Chesapeake Bay. At the time of Mathias's tour Morton was serving as Richard Nixon's secretary of the interior, after stints in Congress and as chair of the Republican National Committee. One of Morton's less well known appointments was to his post on the board of trustees of the Chesapeake Bay Foundation. Morton had encouraged the cofounders of CBF to start the organization in 1966, and he was one of the few holdovers from the original fourteen members of the board a decade later. He would continue to serve on the board until his death in 1979. Although he was not as publicly vocal about the Bay as other politicians, Morton was a critical link between Mathias and CBF and a major behind-the-scenes player in the creation of the organization. Although no record of the conversation exists, during his tour Mathias took a cruise with Arthur Sherwood aboard Sherwood's yacht. Sherwood must have been persuasive, because on February 1, 1974, Mathias officially joined CBF as the newest member of its board of trustees. Mathias would serve continuously on the board until he retired, after which he remained on the board as an honorary member for the rest of his life. As the Chesapeake's recovery became hotly contested at the level of federal policy in the early 1980s, CBF's access to Mathias would be an invaluable avenue for scientists and activists to connect, communicate, influence, and support the Bay in Congress.

A final reason why the legend of Mathias's Bay tour is misleading is that the time between his tour and the subsequent funding of the multiyear, $27 million EPA study gets compressed and oversimplified. It wasn't until July 9, 1974, more than a year after his June 1973 tour, that he released a report proposing the creation of a joint state and federal commission that would oversee planning and resource management. The famous Mathias study was not plan A; rather, it emerged in a somewhat ad hoc fashion as the EPA began to specifically monitor the Bay in response to pressure from Mathias (and perhaps also because of some enlightened self-interest from the EPA administrator, Russell Train, who maintained a summer residence on the shores of the Chesapeake) and other Maryland officials and activists. An EPA study was not what Mathias originally wanted. His initial plan was to shepherd through Congress the creation of a federal planning commission with regulatory powers that would be able to provide top-down coordination and enforcement to the states. In a vacuum, it was an excellent idea, but it stood little chance of ever being implemented. Neither Maryland nor Virginia state legislators had any interest in ceding their authority on Chesapeake matters. Thus, Mathias's laudable efforts to secure multiyear funding for a study to prove how seriously imperiled the Bay was came after his original plan failed to gain traction. The journalist Tom Horton, ever an astute observer of Bay politics, quipped that despite some happy talk, Mathias's original proposal, to create a federal planning commission, "was allowed to die a quiet death."[21]

THE BI-STATE CONFERENCE ON CHESAPEAKE BAY

Mathias's failed bid to create a federal regulatory commission did succeed in encouraging the states to work together by shining a spotlight on the failures of interstate coordination and the duplication of efforts across the watershed. More importantly, Mathias raised the specter of (eventual) federal involvement, which was enough to make Maryland and Virginia take some preliminary steps to cooperate. The most important of these steps was a conference scheduled for spring 1977, which in keeping with tradition bore the utterly uninspired name Bi-State Conference on Chesapeake Bay. Unlike so many previous gatherings, the Bi-State Conference actually resulted in meaningful changes. This result was in part because of the Bay's failure to "bounce back" after Agnes, but it was also a result of a newly invigorated local environmental movement, which in turn drew its strength from the Bay's continued struggles.

As we've seen, prior to Agnes, Chesapeake environmentalism was still in its nascent phase. CBF did not yet have a thousand members by the time of the first Earth Day in 1970. The Citizens Program for the Chesapeake Bay, now the Alliance for the Chesapeake Bay, only began operations in 1971. The success of the national environmental movement in the late 1960s and early 1970s had not created a Chesapeake environmental movement. That only grew in response to changes from the Chesapeake itself. Reflecting on CBF's growth in the late 1970s, its cofounder and director, Arthur Sherwood, said that while the rest of the country was undergoing an environmental awakening "the Chesapeake Bay was pretty well taken for granted." Agnes helped change that. By 1977 CBF had nearly five thousand members, and although its size was a fraction of what it would become, it was already an influential player, with the Bi-State Conference serving as a case in point.

CBF's fingerprints were all over the Bi-State Conference, which allowed CBF insiders to help shape the agenda behind the scenes. Four of the six members of the conference steering committee were current or former CBF board members, and many CBF staff, trustees, and members were among the participants, presenters, and attendees. It is important to understand that the Bi-State Conference was above all a political event, not a scientific one. Although scientists were lead presenters and shared results of their research, the audience was public officials and public opinion, and the goal of the conference organizers was "to provide a forum for interest groups, agency personnel and other persons concerned about the future of the Bay" to begin coordinating a response to the estuary's decline. The most important public officials attending the conference were undoubtedly those representing the EPA. After receiving appropriations to the tune of $25 million to study the Bay, the agency had struggled to implement a coherent research agenda, drawing the ire of Mathias and many other people working on improving the Bay. Although never officially stated as such, the conference's agenda appeared to be designed to help shape what questions the EPA would be asking and answering with its $25 million. The EPA's acting administrator, Alvin Morris, tacitly acknowledged as much in his opening remarks at the conference. He began by acknowledging the "tardiness" of the EPA's program, and he admitted what everyone there already knew, which was that the EPA's "program has not been infused with the 'fire' and speed many of you and we ourselves had wanted." Morris explained that the EPA had had to "draft, re-draft, and re-draft again our proposal and strategy," but he hoped the conference would "aid us significantly in determining the priority needs, establishing the information gaps, obtaining the appropriate work proposals, and funding the programs to acquire the information necessary to make sound decisions." Thus, in the acting administrator's own words, he hoped the conference would help the EPA do its job.[22]

The single most important way that the conference helped shift EPA priorities concerned nutrient pollution. Outside climate change, nutrient pollution is now widely understood as the biggest threat to the Bay, but in 1977 there was not yet a broad consensus about how severely nutrient pollution affected the Bay. When the EPA concluded its study and published its results in 1983, it would definitively identify nutrient pollution as a prime culprit behind the Bay's decline. Yet it was the efforts to highlight the dangers of nutrient pollution at the 1977 conference that helped make the issue a central focus of the EPA's work. At that time, prominent environmentalists like James Gutman, of the Maryland Environmental Trust, argued that sources of pollution from the land were not central to the Bay's decline. Gutman reasoned that the suburbanization of coastal Maryland would lead to less pollution from farms and that the remaining farmers themselves would be incentivized by the "ever increasing costs of fertilizers, herbicides and pesticides ... to have these products 'do their thing' on *their land*—not somewhere else." In his view, concerns about agricultural sources of nutrient pollution were overblown because there would be fewer farmers in Maryland and those that remained would farm more efficiently.[23]

What Gutman did not understand, but scientists and watermen alike were beginning to grasp, was that sources of pollution from upstream were the drivers of the Bay's decline. It is especially notable that Gutman dismissed the threat of *Maryland* farmers, without considering the impact of the vast networks of agricultural lands in Pennsylvania, a state that makes up roughly 35 percent of the Bay's watershed. Watermen like Larry Simns, on the other hand, were quick to blame farmers, in part because of their experience witnessing runoff but also out of self-interest: if pollution from farms was the culprit, then in their view there was less need to strictly regulate the fisheries. Scientists like Dr. J. Kevin Sullivan, director of the Smithsonian's Chesapeake Bay Center for Environmental Studies, had a bit more credibility as objective observers. Sullivan argued that nutrient pollution was causing problems in the Bay's waters but hedged his claim with the classic caution of a scientist discussing the emergence of a new perspective. He noted that "watershed research aimed at establishing relationships between land use, pollutant loading and their effects in the Chesapeake Bay has been underway for only several years. The geographical coverage of much of this work is at present very limited." The work Sullivan alluded to included Operation Agnes and subsequent studies that had only begun as a result of Tropical Storm Agnes. Their preliminary results showed nutrient pollution to be a major factor, but the Bay's scientific community was "just beginning to understand the extent and severity of non-point source loading." [24]

Sullivan's hope, and the hope of many at the conference, was to persuade the EPA to seriously study the issue of nutrient pollution. In that regard, they were highly successful. It is important to remember that a good portion of the conference was political theater—recall acting administrator Morris's public apology for the slow start and his call for help. There were no research surprises, as the conference papers had been circulated at a preconference workshop. Most of the important scientific announcements had been discussed as early as the CRC's 1974 workshop on Agnes research. Politically, the Bi-State Conference organizers succeeded in getting a public commitment from the EPA to take up their research agenda. In response to the conference report on nutrient pollution, the EPA's representative, Robert Moore, said that "manpower positions have been created to address concerns of the workshop report," and he agreed that "the need to emphasize non-point source pollution is apparent."[25]

The EPA would eventually validate the emerging scientific research and observations of watermen in the much more highly publicized follow-up conference in 1983, which concluded with the signing of the first Chesapeake Bay Agreement. As this analysis has shown, the dominant narrative, which begins the story of the Bay's restoration at that pivotal moment in the early 1980s, does not include critical events and actors from the 1970s. It must not be forgotten that key components of the research agenda that culminated in 1983 were agreed to in the less-heralded Bi-State Conference on the Bay in 1977. In other words, by the time the dominant narrative begins, the major action had already taken place, and critical decisions had already been made that would shape the subsequent history of the restoration effort. By starting in the 1980s and focusing on politicians and policy, the dominant narrative utterly misses the work of scientists, watermen, and activists to set the agenda and pressure political leaders to act.

AFTERMATH AND CONSEQUENCES: THE CREATION OF THE CHESAPEAKE BAY COMMISSION

One final result of the Bi-State Conference merits consideration. Unlike so many previous conferences, the Bi-State Conference led directly to a new institution in the Bay's management structure, the Chesapeake Bay

Commission (CBC), which contributed to a higher degree of regional cooperation on Bay policy. An editorial in the Chesapeake Bay Foundation's newsletter suggests that the Bay's leading activists recognized that something had changed at the Bi-State Conference. "The Conference itself appears to have jolted a number of doubters into thinking seriously about the need for a better way of coordinating the activities of Bay related federal, state, and local agencies. That's an accomplishment no other conference has achieved."[26] Although the process would take longer than they might have wanted, the conference led directly to the CBC's inception in 1980, which prefigured the creation of the Chesapeake Bay Program by three years and has served as an important venue for further regional cooperation.

The first concrete step Maryland and Virginia officials took after the Bi-State Conference was to create a joint legislative advisory commission in 1978 to suggest ways that the two states could better coordinate Bay policies. Whereas previous Chesapeake conferences had been dead ends, the Bay's continued decline following Agnes, combined with growing grassroots pressure, forced the two state governments to grudgingly begin working together. The legislative advisory commission considered six possible ways to improve coordination of Bay policies, including creating a state and federal Title II commission, ceding control of Bay policies to the federal government through a variety of mechanisms, creating a states-only organization to explicitly exclude the federal government, and doing nothing. Fortunately, the commission quickly ruled out doing nothing and acknowledged what activists had long complained about, namely, that "there are insufficient mechanisms for policy linkage between the states."[27]

Senator Mathias's preferred outcome was the creation of a Title II commission, which he had been seeking since 1973. Under Title II of the federal Water Resources Planning Act of 1965, the president of the United States could create a joint federal and state river basin commission authorized to make a regional water resources plan, with the commission chairman appointed by the president. Mathias believed that a Title II commission for the Bay would be able to "bring together representatives of the States and the Federal agencies," who could "evaluate the many innovative ideas that have already been suggested to safeguard the Bay."[28] Mathias desperately wanted a mechanism that would allow for coordination and enforcement of Bay policies between the states and the federal government, but as had been the case earlier in the decade, other political figures, especially state legislators, fiercely resisted federal encroachment. Federal monies for research were one thing, but direct involvement in decisions about Bay management was another matter entirely.

It was no surprise when the legislative advisory commission rejected the possibility of an active federal role in Bay management. However, given the long and acrimonious history between Maryland and Virginia, it was surprising that the respective legislatures accepted the group's recommendation that the states work together to create a commission to advise the states on joint policy toward the Bay. Crucially, there was not yet a vision for significant federal involvement. "It is the consensus of the Commission," they wrote, "that the primary responsibility for governing the Chesapeake Bay should remain with the states." While the two state delegations rejected federal involvement and recognized the necessity of cooperation, they were also reluctant to give up either state's freedom of action. "This Commission also finds," they went on, "that it is not necessary or desirable at this time to create an entity having management or regulatory functions." Instead, they determined that "the goal which must be achieved immediately is increased cooperation between the two states themselves." The commission's report made it clear that neither state wanted to yield any of its management agencies' authority. Its proposal for what became the Chesapeake Bay Commission was a fourteen-member body, composed of equal numbers of state legislators from Maryland and Virginia, that would "advise the two legislatures on proposed legislation affecting the use of Bay resources and would serve to focus legislative attention on problems identified by the executive agencies." The new body would be limited in scope; it "would not be involved in any primary research," but most of all, the report emphasized, "the proposed commission would not be assigned any regulatory or management powers." Furthermore, the report reflected some of the conservative rhetoric soon to be popularized by Reagan by defining the commission's role as "an improvement, not as an enlargement of government."[29]

In retrospect, it is understandable why the Chesapeake Bay Commission took the form it did, as an advisory and coordinating commission without regulatory power. At the national level, the political mood had turned against bigger government, and at the local level, each state jealously wanted to guard its prerogatives over the Bay. The commission has repeatedly come under fire because it lacked regulatory authority, and it has often been identified as a failure of Chesapeake environmental politics. For instance, the political scientist Howard Ernst accused the CBC of contributing to a "political dead zone" in the region, in which endless recommendations fail to lead to concrete policy changes, in large part because the CBC lacks any regulatory authority.[30] Indeed, at the time, many of the Bay's leading advocates, like Senator Mathias, were disappointed in the outcome. While some criticisms are valid, they should not obscure the tremendous historical significance of the commission nor the crucial role its members have played in building consensus for environmental policy across the watershed. Since the Great Depression, politicians, scientists, and ordinary citizens concerned about the Bay had been calling in vain for regional cooperation. Again and again, for fifty years Maryland and Virginia had failed to work together to cooperate and coordinate their efforts to manage and protect the Bay. Conference after conference failed to produce a tangible change in approach to the Bay, siloed scientific institutions struggled to coordinate their research efforts, and an uninterested citizenry continued to be mostly apathetic toward the Bay. It took the most devastating tropical storm in the region's recorded history to galvanize fresh scientific research and political pressure inside and outside the halls of power to initiate a process that would eventually force Maryland and Virginia to cooperate. As we will see in the next chapter, once awakened, the Chesapeake's citizen activists would be an important counterweight to the conservative environmental backlash of the 1980s, at least at the local level. Seen in proper historical perspective, the Chesapeake Bay Commission is best understood as a major historical achievement and a sign that a new, regional era of Bay management was underway.

CONCLUSION

Most accounts of the history of the Chesapeake Bay's restoration movement begin with the Chesapeake Bay Program, but as this chapter has argued, starting with Tropical Storm Agnes puts the Bay front and center in the narrative of its recovery while also including a wider range of historical actors. The most important result of starting the narrative earlier is to restore the historical significance of Maryland and Virginia's voluntary agreement to cooperate on Bay issues. Within a few years, the Commonwealth of Pennsylvania would join the Chesapeake Bay Commission and become a full partner in the Chesapeake Bay Program, both of which would have been inconceivable developments in the mid-1970s. The regional approach to saving the Chesapeake began in earnest when Tropical Storm Agnes devastated the watershed. Bay advocates created a new regional framework that, while not ideal, has allowed for the movement to grow, incorporate new interests, and sustain involvement in improving the Bay for more than four decades.

It is important to stress how soon after Tropical Storm Agnes hit the region a consensus emerged about what ailed the Bay. While much research still was needed to understand precisely the reasons for the Bay's decline, the overarching challenge was to present knowledge about the Bay's peril in such a way as to produce the political will to act. The next chapter takes up this story in full, so a few concluding remarks here will suffice. Larry Simns provides an interesting coda to this whole period. He remembered that when the multimillion-dollar EPA study finally produced its report, he "received a direct call from Senator Mathias. He informed me of the report and that the recommendations were done and that the watermen of Maryland were spot-on with their assessment. He was shocked how simple watermen like me could see the problems of the Chesapeake Bay so clearly."[31] From Simns's perspective, he and his fellow watermen had had it all figured out by 1972, but it had taken the rest of the world a decade and tens of millions of dollars to catch on. Simns's recollection should be taken with a grain of salt, but it still makes an important point. To those who knew the Bay the best, its problems were obvious by the middle of the 1970s. As this chapter has shown, many previous attempts to do something about the Bay had failed miserably, and it took a tremendous effort to generate a political response to the Bay's ecological crisis. Starting the narrative with that political response fundamentally misses the early drama of the Bay's homegrown environmental movement. Without dismissing valid observations about the flaws and the failures to fully realize a healthy Chesapeake, by beginning our narrative with Agnes, we can appreciate that it was still a major accomplishment to mobilize and create a new political structure after a half century of false starts and dead ends. While messy and incomplete, the Bay's halting recovery since the depths of the mid-1970s represents not a failure of restoration but a success of sustainability.

THREE
THE CHESAPEAKE BAY AGREEMENT

The text of the original Chesapeake Bay Agreement fits comfortably on a single piece of standard, 8½" x 11" paper. Despite its modest dimensions, this document represented a turning point in the Chesapeake Bay's environmental history. It set the pathway for regional cooperation over the Bay for the subsequent generation of environmentalists, scientists, and politicians. As implied by its name, this was a voluntary *agreement* between state and federal partners that acknowledged "an historical decline in the living resources of the Chesapeake Bay" and committed to "share the responsibility for management decisions and resources regarding the high priority issues of the Chesapeake Bay." It was the first time that all the region's political officials, notably including those from Pennsylvania, and the federal government recognized that the Chesapeake was in trouble and agreed to do something about it, however modest.[1]

As the preceding chapter argued, this was not the beginning of the Chesapeake Bay restoration movement, but it was a crucial turning point that favored a particular approach to solving the Chesapeake's problems. Some scholars have criticized the lack of enforcement mechanisms in the first Chesapeake Bay Agreement and its successor agreements, arguing that the system it helped inaugurate "is better suited to resolve conflict than it is to produce tangible environmental outcomes." While that and similar criticisms have merit, the historical context in which stakeholders made these agreements not only explains why events unfolded the way they did—conflict resolution is actually quite an important feature of the system—but allows us to better appreciate why this humble document was one of the major turning points in the Chesapeake's environmental history and how it has contributed to the politics of saving the Bay. Because of its shortcomings, it is easy to take the Chesapeake Bay Agreement for granted. However, the story of its inception has as many plot twists, heroes, and villains as a Hollywood blockbuster, with a surprise ending that illuminates the magnitude of the achievement.[2]

A CHANGING ZEITGEIST

One of the greatest things that can happen to a historian poking about in the archives is to stumble across a document that perfectly captures a transitional moment in history. It can be unpublished works by an unknown genius, a famous treaty or law seen in a new light, or in the case of the Chesapeake Bay something more prosaic: a piece of hate mail. In early December 1980 the Chesapeake Bay Foundation received a vitriolic letter from a man named Donald Deborde. Deborde was one of the tens of thousands of people contacted by CBF during the year-end membership drive. Unfortunately for CBF, it would not receive any membership dues from him, but fortunately for the historian, CBF reprinted Deborde's missive in their own newsletter. This letter provides a window into the backlash to environmentalism that was emerging in the region and around the country. Why was Deborde so angry? Let us hear from him in his own words:

Dear Do Gooders;

Let me get this straight! You want me to donate so that you can:

1. "Educate" more environmental nuts like yourself;
2. lobby for more laws and regulations and restrictions so that the bay is impossible to use;
3. buy more land for the ducks

Nuts to you! The bay and damn ducks are doing just fine. It's the boaters, fisherman, and property owners who need help. So tell me, where do I sign up and donate to fight you jerks? I bet you are Democrats. Me and Ronnie are going to take care of you.

Disgusted,

Donald D. Deborde

This letter is interesting for a number of reasons. For instance, Deborde would have lost whatever money he wagered on CBF being Democrats; as discussed, the founding group included a number of prominent Republican politicians, and as detailed in the previous chapter, one of the Bay's greatest champions in Congress was the Republican senator from Maryland, Charles Mathias. The suggestion that the Bay was doing "just fine" was in stark contrast to the observations of concerned watermen and the conclusions of the region's scientists. But these are not what make Deborde's letter so compelling to the historian. Deborde also mentions "Ronnie," which was likely a reference not to a local tough guy who could really mess you up but rather to President-elect Ronald Reagan.[3]

Deborde's threat to "take care of" CBF with the president-elect's assistance would have made sense to contemporary followers of Reagan's presidential campaign or even to modern readers with a general familiarity with the fortieth U.S. president's well-deserved reputation for rolling back governmental interference in the lives of liberty-loving Americans, particularly with respect to environmental laws and regulations. Fortunately for CBF and the wider Chesapeake environmental movement, Deborde's threat would not come to pass. One of the most fascinating aspects of studying history is that there are often deep ironies when it comes to historical actors' expectations for the future. Reagan's effect on the Chesapeake was not the greatest historical irony, but it certainly was not the least. While Reagan was no great friend to the environmental movement nationally or in the Chesapeake, he did reluctantly end up supporting the Chesapeake Bay environmental movement on all the points Deborde complained about. Imagine the look on Deborde's face as he watched the 1984 State of the Union address and heard The Great Communicator proclaim his personal commitment to supporting "the long, necessary effort to clean up a productive recreational area and a special national resource—the Chesapeake Bay."[4]

This detailed reading of Deborde's letter highlights a key insight from the history of Chesapeake Bay environmentalism that is a central argument of this book: the movement played out with different sets of actors and with a different chronology than are typically ascribed by environmental historians. This is further evidence for one of this book's main points, namely, that a multitude of American environmentalisms emerged in the latter decades of the twentieth century, not a monolithic movement. As discussed, the 1960s and 1970s were not a time of a great environmental awakening in Chesapeake environmentalism. It was not until the end of the 1970s that Chesapeake environmentalism began to take on the shape of a movement with widespread political support capable of pressuring elected leaders for change. By the 1980s, it was even capable of forcing an antienvironmental, antigovernment crusader such as Reagan to change his tune, at least in one small area. However, the strength of the movement was far from clear at the time Deborde wrote his letter. In 1980, with Reagan just about to take office, it would have been entirely plausible to assume that Reagan's election spelled imminent doom for the nascent Chesapeake environmentalism; indeed, that was very nearly how events unfolded. Instead the exact opposite happened. Unpacking this story requires us to understand the state-level dynamics of Maryland and Virginia, the national politics of the Reagan administration, and the ongoing ecological turmoil of the Bay, but most of all it highlights the dedication of a committed group of activists and leaders who refused to let an opportunity to change history slip away.

REAGAN AND THE ENVIRONMENT

When Ronald Reagan took office in January 1981, he famously declared that "government is not the solution to our problem; government is the problem."[5] Historians are just beginning to explore the consequences of the Reagan era (and, it must be noted, they are continuing to unfold), but even so, his effect on U.S. environmental movements seems curiously understudied. Perhaps because of his administration's widely recognized hostility toward environmental protection, there has been an assumption that the effects were straightforwardly negative. A 2004 study found that while "a number of observers have already evaluated Reagan and suggested his place in history, his environmental record receives little attention."[6] The situation has changed little since 2004, making this analysis of Reagan and the Chesapeake Bay all the more important to the ongoing attempt to understand and describe the complex, at times contradictory effects of his presidency on environmental movements in the United States.[7]

A brief, big-picture overview of the Reagan years reveals that his administration accelerated the polarization on environmental issues that would feature so largely in U.S. politics in the late twentieth and early twenty-first centuries. At the national level, pro- and antienvironmentalists reacted to the Reagan administration's overt hostility toward environmental protection. For environmentalists, this resulted in a surge of support for their cause. Moderate Republicans opposed Reagan in Congress, individuals joined or donated to environmental groups, and a wholesale dismantling of environmental laws was largely avoided. For environmental opponents, the reaction was to only dig deeper and double down on attacks against regulatory overreach while pushing the Republican Party further to the right so that politicians like Charles Mathias, once a dark-horse presidential candidate, no longer had a home in the party.[8] Environmental policy in Reagan's first term was highlighted by national controversy at the EPA and in the Department of Interior, where the names of his political appointees, Anne Gorsuch[9] and James Watt, became synonymous with antienvironmentalism and bureaucratic ineptitude. Reagan's attempts to undermine environmental regulations went further than controversial appointments; they included executive orders and massive budget cuts that were part of his administration's larger program to reduce federal spending. Reagan ultimately failed to repeal any major environmental legislation because of opposition from Congress, including many members of his own party, and from environmentalists, who successfully rallied public opinion in defense of the environmental legislation of the 1970s. Reagan found more success, however, using the administrative powers of the Oval Office to influence, and in most cases undermine, environmental policy. Reagan's agenda—and more importantly, its clumsy implementation by his appointees—led to a fierce backlash that forced Reagan to replace Gorsuch and Watt in 1983 in order "to defuse the issue."[10]

For most of his second term, in the words of Judith Layzer, "Reagan paid scant attention to the environment" as his new appointees implemented his policies more quietly, and a proenvironment Congress passed legislation advancing environmental protection. Notably, Congress reauthorized the Clean Water Act over two Reagan vetoes, increased appropriations for the Superfund program, renewed and strengthened the Safe Drinking Water Act (by a 94–0 vote in the Senate), and reauthorized the Endangered Species Act despite vigorous opposition from the White House. As an indication of how far the Republican Party has moved away from environmental protection since the 1980s, Reagan also signed the Global Climate Protection Act and designated 10 million acres of federal land as wilderness. However, scholarly opinion is that despite some notable achievements, "on balance, Reagan's administrative presidency strategy had a blighting impact on environmental and natural resources policies." This may have been true at the nation level, although it is unclear whether the negative impacts were specifically from the Reagan administration or from subsequent antienvironmental shifts by Republicans, but the picture is much more complicated at the regional level. As we will see, on balance the Reagan administration ended up aiding the cause of Chesapeake environmental protection, though that certainly was not the original intention.[11]

The Reagan administration's most direct impact on Chesapeake environmental politics came from the EPA's meltdown during Reagan's first term, under Anne Gorsuch. An early indication that there would be problems at the EPA was the administration's lengthy delay in nominating Gorsuch. The U.S. Senate eventually confirmed Gorsuch as EPA administrator in May 1981, more than three months into Reagan's term. She had no experience in Washington, which was a bit unusual at the time, and her environmental background was limited to ardent opposition to the EPA as a member of the extreme right wing of the Colorado state legislature, who dubbed themselves the "House Crazies." Her main qualification for the job was devout loyalty to Ronald Reagan. Gorsuch eventually became the first agency head in U.S. history to be cited for contempt of Congress, and she resigned in disgrace in March 1983, ending a period of "two-and-a-half years of near chaos" at the EPA. Gorsuch's main goal at the EPA was to cut the agency's budget and downsize its staff. This initiative fit well within Reagan's larger political agenda, which "centered on the budget process, curbing budget requests through Reagan loyalist agency heads and OMB." The EPA's budget declined by 44 percent during Reagan's first term, which led in part to a 29 percent decline in the number of full-time employees at the EPA. Budget cuts were not the only reason for the loss of talent within the agency, as "Gorsuch's open contempt for the EPA bureaucracy led many seasoned employees to quit." In all, more than four thousand EPA employees resigned, leading former EPA chief Russell Train to publicly castigate Gorsuch. In a *Washington Post* editorial he wrote that "it is hard to imagine any business manager consciously undertaking such a personnel policy unless its purpose was to destroy the enterprise." Gorsuch aroused public antipathy almost as soon as she left the Senate confirmation hearings, yet she would remain at her post until a scandal surrounding her refusal to release EPA documents pertaining to the handling of $1.6 billion in Superfund monies created intolerable political fallout for the Reagan White House. She was replaced by the respected former EPA head William Ruckelshaus, but "in both its substance and its tactics, Reagan's deregulatory initiative caused deep and lasting damage to EPA."[12]

The EPA's Chesapeake Bay Program is a superb example of what the dysfunction of Reagan's environmental policy looked like. In 1981, after years of research, conferences, and more research, the EPA had nearly completed its report on the Chesapeake's environmental health but needed another $3 million to complete its existing projects and publish its findings. Unsurprisingly, Gorsuch had no interest in ensuring that the agency finished the job. One of her first actions after being confirmed in May was to announce in June that the EPA only sought a $1 million outlay to conclude the study, but, crucially, not to publish any of the results. This was in keeping with established policy from the Reagan White House. In April 1981 Reagan had issued a moratorium on all new publications by the federal government, and "OMB immediately issued a bulletin confirming the moratorium and ordered government departments to undertake a comprehensive review of existing and planned publications in order to eliminate or reduce duplicative, wasteful, and otherwise 'unnecessary' activity." In Gorsuch's eyes, disseminating the results of more than a half decade's research at a cost in excess of $25 million was wasteful and unnecessary. Understandably, Chesapeake environmentalists thought otherwise. Under the headline "EPA Program Axed," the Chesapeake Bay Foundation warned its members that if the "budget recommendations are carried out ... the complete findings will simply be printed on microfiche sheets for storage. They will not be fully developed into meaningful conclusions and recommendations for the benefit of policy-makers and planners as intended." Such a mildly worded protest surprised many of CBF's own members, given the severity of the threat to the Bay Program. In 1981 CBF, like the Chesapeake's environmental movement as a whole, was in the middle of a transition, and it was struggling with how to react to the Reagan administration's blatant disregard of environmental issues. In response to challenges from the federal government, CBF and other environmentalists would ultimately retool their strategies, grow stronger, and succeed in making the president care—or at least care about appearing to care—about the Chesapeake Bay.[13]

FROM CONTRIBUTING TO LEADING

In response to the federal government's virtual abdication of its responsibility to protect the environment, CBF's executive officers and senior staff began to rethink the foundation's role in the Chesapeake environmental movement. Former CBF vice president Don Baugh described this transition as one "from contributing to leading." What he meant was that beginning in the early 1980s, he and other senior CBF staff began to see the foundation's role differently than they had in the 1970s. Baugh and others thought CBF needed to play a larger regional role and become a more forceful advocate for the Bay in order to galvanize the popular support necessary to get lawmakers to put policies in place that would restore the Bay. Although Baugh was careful to explain that it "was a transition that took time and it was not a clear transition," documents from 1981–82—a period coinciding with increasing friction between CBF and the federal executive branch—show a distinct shift in CBF's philosophy, strategy, and leadership. One of the key developments for both CBF and the Bay restoration movement was the rise to prominence of William Baker as the organization's leader. Baker had joined CBF as an intern in 1976 and quickly risen to become the foundation's second-in-command. However, when Arthur Sherwood stepped down at the end of 1979, CBF's trustees chose an older hand with connections in DC, David McGrath, to succeed him. McGrath lasted little over a year before being forced to resign after repeatedly clashing with CBF's board. The second time around, Baker got the top gig and proceeded to transform CBF over his four decades in charge. As we will see, this combination of microevents within CBF and macroevents within U.S. politics led to the emergence of Chesapeake environmentalism as a serious political force.[14]

One of the fundamental differences between Baker's leadership and that of CBF's old guard was a more aggressive approach to protecting the Bay, though this approach emerged too slowly and was not aggressive enough for some critics. Since ousting Jess Malcolm in 1971, Sherwood had piloted a careful and cautious course for the organization throughout the 1970s, focusing on environmental education and the enforcement of existing laws and regulations. Sherwood's singular achievement may have been to lay the foundation for CBF's environmental education program, which Baugh built upon and made a backbone of the organization. Education suited Sherwood's core philosophy, which he called "environmental reasonableness." Sherwood believed that all uses of the Bay could be accommodated if people were "reasonable" and accepted commonsense compromises. Oddly for an environmentalist, he rejected strictly ecological arguments in favor of the Bay. In the parlance of an earlier time period, Sherwood was a conservationist, not a preservationist. He described his approach as "a course better understood as plodding than inspired, cautious than thrusting, doubting than assured, conservative than avant garde, respectful of other's opinions than confidently assertive."[15] When Sherwood stepped down, his philosophy was already being challenged by some inside the organization who thought his approach too conciliatory, but it was the intransigence of the Reagan administration in particular that made a policy of "reasonableness" increasingly untenable.

At the start of Reagan's first term, CBF, with Baker at the helm, remained committed to a more moderate approach to environmental policy. An editorial response to a member's letter from the June 1981 issue of *CBF News* (the same issue that contained CBF's rather mild protests about Gorsuch's plans for the EPA) illustrates the foundation's position at the time. The member, a self-described "radical environmentalist," was "on the verge of giving up on CBF" because it was not doing enough to defend the Bay and was "almost apologetic when criticizing despoilers." CBF's response, likely penned by Baker himself, was that "CBF was founded on the premise that the Bay can support many diverse uses, each in its proper context and with appropriate safeguards, without compromise to its renewable natural resources.... Our philosophy is articulated in the CBF charter's statement of goals: to wit, 'to contribute to the wise management of the natural resources of the Chesapeake Bay.'" This editorial continued to lecture the "radical" member that "in a democratic government, law represents a consensus of the will of the people . . . so it's a fundamental principle with CBF to work within the system and not attempt to force solutions using methods which are apt to be destructive to the rule of law and even to democracy itself." Simply put, CBF was reluctant to be highly critical of the Reagan/Gorsuch EPA for fear of being identified with what it saw as the "radical" wing of the environmental movement. Baker would reaffirm the organization's commitment to contribute to the wise management of the Bay's natural resources in his first annual report in 1981, though he and the organization would soon step out of Sherwood's shadow.[16]

A year later CBF's tone had changed dramatically. In his cover letter to the 1982 *Annual Report* the chairman of CBF's board wrote that it was no longer enough for the organization to be part of the Bay's supporting cast. Instead, "CBF must be at the center of efforts to protect the Chesapeake Bay" and must "take a leadership position in the formulation of management recommendations and implementation strategies." This shift was in part a result of increasing frustration with the Reagan administration. Baker wrote in a report to members that "with the drastic reduction in federal involvement in Bay affairs our job is now more important than ever." He complained that "political pressures" had "force[d] dozens of agencies to compromise their efforts to maintain environmental safeguards," adding that "it becomes increasingly clear that many of our present activities simply serve to bolster stopgap measures for Bay protection." Baker was ready for CBF to take on a larger role in advocating for the Bay, and he believed the organization could succeed. Baker noted optimistically that "the last year has shown a new groundswell of concern for the Chesapeake Bay. We believe that our mandate for protecting the Bay has never been as strong and that we have the constituency to put muscle behind our words." Indeed, CBF saw its membership nearly double from roughly 7,800 in 1980 to over 15,000 in 1983. The organization benefitted from events as much as it drove them, absorbing members who wanted to join the fight to "save the bay" and resist the Reagan administration's efforts to crush the Chesapeake Bay Program. In short,

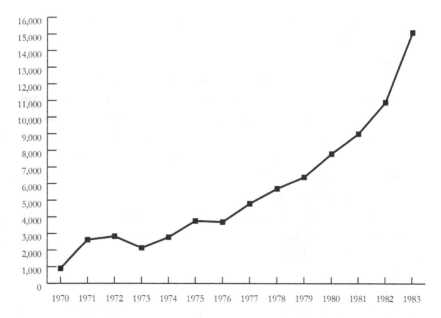

Chesapeake Bay Foundation membership, 1970–1983. (*Source:* Annual reports, CBF internal records, Philip Merrill Environmental Center, Annapolis, MD)

during Reagan's first term CBF leadership saw the need for the organization to be more assertive and its potential to do so.[17]

Other examples of CBF's shift in 1982 come from some of the organization's front-line activists. Stuart Lehman was one of CBF's two staff biologists. He joined CBF in 1979 while working on a master's degree in environmental science at George Washington University, and he ended up working on one of CBF's major pollution-enforcement programs for a decade. In 1979 much of Lehman's work involved close collaboration with federal agencies. He wrote proudly that he had worked with officials in the EPA and the U.S. Fish and Wildlife Service "to set up federal channels to coordinate our joint comments" on pollution permits. By 1982 the situation had changed substantially. Lehman said that he was "concerned that due to recent budget cutbacks and staff reductions, the state and federal agencies will not be able to do an adequate job enforcing the law." Furthermore, he thought that CBF needed to be prepared to "step in and see that the appropriate action is taken" because of the federal government's failure to fulfill its oversight responsibilities. Previously Lehman had reviewed pollution permits, often in collaboration with other agencies, to try to set limits on how much of a given pollutant could be discharged into the Bay. When Lehman believed permits were too lenient, he would call in CBF's staff attorneys and/or pro bono legal volunteers to help push for tighter restrictions. Crucially, once the permits were issued, Lehman and CBF typically left the monitoring and enforcement of those permits to state and federal agencies. By 1982 it was clear to Lehman that this system of relying on enforcement by federal agencies was not working, and the program he led shifted its focus from reviewing new permits to pursuing independent enforcement of violations to existing permits. Put differently, CBF started doing the regulatory job that government agencies had abdicated. In a few years' time, this new focus would lead CBF's attorneys all the way to the Supreme Court in one of the foundation's highest-profile legal battles, *Gwaltney of Smithfield, Ltd. v. Chesapeake Bay Foundation, Inc.* But from the perspective of 1982, it was yet another indicator that the organization was changing its approach to saving the Bay, and it confirms Don Baugh's recollections that the early 1980s were a period of significant transition for the organization.[18]

These changes in CBF's approach reflected a broader shift in national and Chesapeake Bay environmentalism. Nationally, Reagan underestimated the level of public support for environmental protection, and "to conservatives' chagrin, the administration's direct attacks on environmental regulation reinvigorated the environmental movement and provoked a backlash among the public. The public's concern redounded to the benefit of environmental groups, which saw immediate financial benefit: between 1980 and 1985, the budgets of the Sierra Club, the National Audubon Society, and the Wilderness Society more than doubled." CBF's experience mirrored that of the national groups, as its budget tripled and its membership quadrupled over the same time, tapping into a surge of public support for protecting the Chesapeake. This public support for environmental protection strengthened the resolve of many in Congress who were inclined to oppose Reagan. Indeed, scholars have made the obvious point that "the greatest obstacle in Reagan's path was his lack of Congressional control." Not only did Democrats retain a majority in the House but moderate and liberal Senate Republicans opposed many of Reagan's goals. Mathias, who earlier in his Senate tenure had supported civil rights bills and troop reductions in Vietnam, was on CBF's board of trustees and worked closely with the foundation leadership to coordinate their policy toward the Bay. Other moderate GOP senators, such as Robert Stafford of Vermont and John Chafee of Rhode Island, who had strong environmental records and key leadership positions on the Senate Committee on Environment and Public Works, were opposed to rolling back environmental protection. Mathias and his Democratic counterpart in the Maryland delegation, Paul Sarbanes, were thus able to accomplish a classic bit of negotiation and horse-trading and split the difference in proposed funding for the EPA's Chesapeake Bay study, succeeding in securing an appropriation of $2 million to finish the study and publish the results.[19]

INITIAL RESULTS FROM THE EPA STUDY

After six years and more than $27 million spent, the EPA finally began to release reports that confirmed what most Bay observers already knew: the estuary was at risk of near terminal decline, and without swift action a systemwide collapse was possible. Although the general outlines were already known, the EPA's report still carried great scientific weight and prestige, despite Gorsuch's efforts to tarnish the agency's reputation. The EPA's findings were released in three phases over the course of a year, beginning in late 1982 with the release of "Chesapeake Bay Program Technical Studies: A Synthesis," which was a brutally dense, 632-page technical summary of the EPA's research findings. This report was not designed to be easily accessible or easily understood by the general public, though it still generated waves when it was released. Reflecting many years later, William Baker was convinced that Anne Gorsuch had done her best "to bury that report" by releasing a massive file of technical details and delaying the compilation and release of any policy recommendations based on the EPA's findings.[20] After Gorsuch resigned in disgrace in March 1983, Reagan would appoint William D. Ruckelshaus to restore some credibility to the EPA and soften his administration's antienvironmental reputation before his reelection campaign got underway. Under more amenable leadership, the EPA would release its final report on policy recommendations in September and then join with the governors of Maryland, Virginia, and Pennsylvania to sign the first Chesapeake Bay Agreement that December.

Gorsuch's effort to kill the EPA's final report through bureaucratic slow walking got off to a good start. When the EPA released its technical report in 1982, it made some local headlines but did not garner any broader national news. For instance, the *New York Times* did not mention any of the EPA's findings from the technical report, though it would report on the final report in 1983. Still, to those paying close attention the report confirmed many of their worst fears and colored their approach to activism. John Page Williams had joined CBF in 1972 to expand its education program, and by 1982 he was running an operation that took roughly fifteen thousand participants on field trips across Maryland and Virginia. Williams spoke for many at CBF when he wrote that "it has been our aim for 10 years to offer field trips that help participants both to understand and to appreciate the Bay . . . but 1982 brought a sense that understanding and appreciation are not enough." Like so many others, Williams saw 1982 as a turning point. According to him, the EPA's technical summary "had a pervasive effect on CBF." Williams explained that "since CBF started offering field trips on a full-time basis, all of us who work in the field from day to day have been aware of problems" ranging from siltation in creeks, habitat loss, fisheries closed for toxic or sewage contamination, "all [of which] form an ominous backdrop to what we see that is alive and healthy." While individual experiences could be dismissed as mere anecdote, the EPA's report "confirmed apprehensions about environmental threats with broad-scale scientific diagnosis." Just as other aspects of CBF's operation adapted during 1982, so too did the education program. Williams wrote that in addition to inspiring students with the magic of the great estuary, the EPA's report "challenged us to build non-point source pollution, nutrients overenrichment, oxygen depletion, and toxic chemicals into our trip curricula as major study areas and not just secondary topics for discussion." Like others at CBF, the education staff would have to "strike some harsher notes too."[21]

While local activists like those at the Chesapeake Bay Foundation were striking harsher notes, at the national level the U.S. Congress was about to make a historically harsh pronouncement. On December 17, 1982, the House of Representatives voted 259–105 to cite Anne Gorsuch for contempt of Congress for failing to turn over documents related to the administration of the $1.6 billion Superfund toxic-waste program. Notably, fifty-five Republicans crossed party lines to vote against Gorsuch. It is important to remember that even as late as the 1980s, environmental issues held broadly bipartisan appeal; only recently have they become victims to the hyperpolarization of American politics. However, the most historic aspect of this vote was not its bipartisanship but rather that it gave Gorsuch the dubious distinction of being the first cabinet-level official ever to be cited for contempt of Congress. Despite the various scandals in U.S. history, from Teapot Dome through Watergate, a cabinet official had never been cited for contempt of Congress. Gorsuch's reaction was characteristically as flippant as it was defiant. She said that the vote "wouldn't do anything," that she would continue to refuse to turn over documents, and that Congress could "send me to jail for a much-needed rest." The EPA had previously promised to follow up its 1982 technical synthesis with another report by January 30, 1983. Predictably, as the calendar rolled into 1983, Gorsuch's EPA was in no hurry to release the next phase of its report on the Chesapeake Bay, and the deadline came and went without a report. This was a "devastating time" for the EPA, which, in the words of one scholar, "experienced unprecedented levels of disorganization, demoralization, and internal strife." The EPA's effectiveness decreased dramatically, even by 1981 or 1982 standards, as "EPA's leaders became isolated from their staffs and increasingly uninvolved in matters not pertaining to the [Superfund] documents controversy."[22]

As the weeks passed without any word from the EPA or any resolution on Gorsuch's contempt charges, Maryland's Senator Charles Mathias scheduled a committee hearing on March 1, 1983, to turn up the heat on one of Gorsuch's underlings responsible for the delay, Region III Administrator Peter N. Bibko. Mathias did this in close coordination with CBF, summoning William Baker to testify, as well as collaborating with Maryland's other senator, the Democrat Paul Sarbanes. Before grilling Bibko about the EPA's delays, Baker testified first to set the stage. In something of a slight, it was Baker who summarized the EPA's findings, rather than Bibko, who did not get the benefit of reporting major findings, only the burden of defending his unit's shortcomings. Baker did praise the scientific research the EPA had done over the preceding years in his opening remarks, but he pointedly called out the failures of the agency's management recommendations. "CBF has reviewed the only preliminary draft of the management report," he said, and "we did not give it high marks." Baker expressed some optimism, highlighting the efforts of program staff, but he had no kind words for the current leadership. Without mentioning Bibko by

William Baker testifying before Senate Subcommittee on Governmental Efficiency and the District of Columbia, 1983. *Front row from left*, Charles H. Collins, CBF assistant director; Richard R. Gardner, CBF general counsel; Baker; Carl Osborne, CBF staff scientist. (Courtesy of Chesapeake Bay Foundation)

name, Baker said that under the previous regional administrator the program had been "superbly administered" but that afterwards there had been "confusion and ineffective management." Baker also accused the EPA of failing to heed the lessons of its own research, stating that "we wonder what is the point of spending $27 million if the findings are to be ignored."[23]

After his opening remarks, Baker answered questions, some of which were about the report but many of which turned to the question of what to do after the EPA's report finally came out, in other words, how to implement the EPA's recommendations. Baker raised a key concern, namely, that "the challenge of having Maryland, the District of Columbia, Virginia, and Pennsylvania work together as they have never done before is substantial." From the perspective of the present day, it can be hard to remember how difficult it was—and still is—to get the states in the Chesapeake's watershed to coordinate their policies. Another challenge was funding the cost of implementing the management recommendations, which were already widely expected to grossly exceed the costs of the EPA's research, which was expensive enough.

Baker feared that even if Mathias and his congressional allies could come up with the appropriations, without some plan to coordinate action "it would be all too easy for the States . . . to try and get parts of Federal moneys and to go their own separate ways without coordination."

After Baker's testimony it was Peter Bibko's turn to defend the EPA's delays in reporting and its role in Chesapeake Bay protection. It did not go well for Bibko. Bibko reviewed the EPA's work in his opening statement and claimed that the agency was on track to meet a September 30, 1983, deadline to conclude the program and release its management report. Unfortunately for him, Senator Mathias was not reassured. He asked Bibko, "Who is making policy? What is the first level of policymaking as far as the bay program is concerned?" Essentially, Mathias was demanding to know who was in charge and who was responsible for the program's delays. To his credit, Bibko did not dodge the question. "I am directing the policy of the Bay program," he replied, but after a back-and-forth about staffing and personnel, Mathias told Bibko that he was "concerned" about his answers and his plans for finishing the report. He said that he would be disappointed if there was any further delay "because this committee expected it in January. As far as we are concerned, it is really overdue now." One can imagine Bibko squirming under Mathias's glare as he replied, "I agree 100 per cent that you are right." However, Mathias wasn't finished. He proceeded to rake Bibko over the coals for contradictory remarks he had made about the EPA's role in Chesapeake affairs after the study phase of the Bay Program ended. After getting Bibko to agree that the EPA would need to maintain a liaison office in Annapolis, Mathias stated sarcastically, "Looking at the 1984 budget, I don't see this reflected. I am hoping it is buried down somewhere in the budget provision for this." Bibko replied vaguely that the EPA would work with the states to help implement new policies, "and we intend to supply the manpower needed to fulfill that obligation." Mathias continued trying to get a more concrete commitment from Bibko, culminating in this exchange:

> **MR. BIBKO:** Senator, I can assure you, please, let me personally assure you on behalf of the agency that the Chesapeake Bay Program will be timley [sic] finished and that we will work very closely with the States to make certain that the fruits of the program are implemented.

SENATOR MATHIAS: So the end of the research program certainly is not the end of EPA responsibility for the Chesapeake Bay.
MR. BIBKO: That is correct, sir. It certainly is not.
SENATOR MATHIAS: And there is a continuous obligation to implement clean water strategies.
MR. BIBKO: There certainly is, sir.

After that exchange, Mathias called for a brief recess, but Bibko wasn't quite out of the hot seat yet.

After the recess, Mathias's counterpart in the Maryland delegation, Paul Sarbanes, took over the questioning. Unlike Mathias, Sarbanes did not have a reputation as a leader on environmental issues. Nevertheless, as one would expect from someone who ended up winning election to the Senate five times and serving for thirty years, Sarbanes had an excellent sense of political currents, especially in his home state, and he was not about to pass up an opportunity to score some easy political points. His interrogation of Bibko made Mathias's cross-examination look tame. Moreover, as a Democrat, Sarbanes was under much less pressure to play nice with the Republican administration. The following example is typical of their exchanges, with Sarbanes repeatedly interrupting Bibko to drive home his point:

SENATOR SARBANES: Are you doing the annual water quality study this year in the bay?
MR. BIBKO: This year we are beginning the first step in establishing a coordinated water quality and perhaps biological monitoring program that will be needed by the States in the years to come. We have taken the money—
SENATOR SARBANES: That wasn't my question, if I am following that answer carefully, and I think I am. My question is, is EPA doing a water quality study this year?
MR. BIBKO: To my knowledge Senator—
SENATOR SARBANES: The Chesapeake doesn't stop, you know, when your research stops. It continues. It is an ongoing organism.
MR. BIBKO: Senator, may I—
SENATOR SARBANES: What are you doing to stay abreast of it?

At this point, Bibko couldn't take it any longer and tried to call in one of the Bay Program project managers to answer Sarbanes's questions. Unfortunately for Bibko, Sarbanes was not about to let him off the hook. Rather than letting another witness take questions, Sarbanes asked pointedly, "I understood you to respond to Senator Mathias earlier that you are the policymaker on this matter. Is that correct? I think he asked you a question about that and you said, 'I am the policymaker.'" The one-two punch of Mathias and Sarbanes would eventually leave Bibko so turned around that he had to ask Sarbanes to repeat his question. Like Mathias, Sarbanes wanted to ensure that the EPA finished its report and maintained a key role in Chesapeake environmental policy after the research phase of the Bay Program ended. The two were determined to extract a commitment from Bibko to keep the EPA engaged with the next phase of the Chesapeake's recovery, but that simply wasn't something the Gorsuch EPA was willing to do. As Sarbanes told Bibko toward the conclusion of the hearing, "You have a chance to make a great contribution, Mr. Bibko, with respect to Chesapeake Bay. We are just trying to get you to seize it here."

It was not Peter Bibko's fate to make a great contribution to the Chesapeake Bay. On March 9, 1983, just a little over a week after his heated testimony, Anne Gorsuch suddenly resigned from her post as EPA administrator. Her leadership at the EPA had created an intolerable level of criticism of the Reagan administration, and Reagan's chief of staff, James Baker, was urging his boss to fire her if she did not resign. In her resignation, Gorsuch laughably claimed that it had been congressional interference that "crippled my agency," not her own incompetence and hostility toward the EPA's fundamental mission. For his part, Reagan told her that he believed she had been "unjustly attacked," but it seems likely that he was privately relieved to be rid of the distraction. Peter Bibko's turn to leave came just two months later. Gorsuch's abrupt resignation caught many senior officials off guard, but they soon followed her out the door. Others were asked to resign after Reagan appointed a new EPA administrator. Bibko and several other senior officials fell into this latter camp and were forced to resign on May 21, 1983. One is tempted to wonder how things might have turned out differently for Bibko if he had seized the moment to make a great contribution to the Chesapeake Bay. As it turned out, the man who asked for Bibko's resignation, William D. Ruckelshaus, would be named Conservationist of the Year for his role in rescuing the EPA's Chesapeake Bay Program from oblivion and signing the Chesapeake Bay Agreement. It would be Ruckelshaus, not Bibko, who made history by committing the federal government, via the EPA, to a central role in a coordinated, regional approach with the states to protecting the nation's mightiest estuary.[24]

CHOICES FOR THE CHESAPEAKE

In the aftermath of Gorsuch's departure, Reagan was eager to turn a liability into an asset. For that reason, he turned to William Ruckelshaus, who had been nominated by Richard Nixon and served from 1970 to 1973 as the first EPA administrator. Ruckelshaus appeared to genuinely care about the EPA's role in protecting the Chesapeake, and with strong grassroots support from organizations like CBF, as well as support from congressional allies like Mathias and Sarbanes, he helped lead the way to the Chesapeake Bay Agreement. However, one of the key players wasn't yet on the board: the Commonwealth of Pennsylvania. In its first two decades Chesapeake environmentalism had been a Maryland and Virginia affair, and as the previous chapter demonstrated, getting those two states to cooperate had been a challenge. Adding to the mix Pennsylvania, home to more than a third of the Bay's watershed but none of the estuary's waters, would be essential to successfully sustaining the Chesapeake. The story of how Pennsylvania became a Bay state further underscores that the 1983 Chesapeake Bay Agreement was a turning point and why the flexible management structure of the Chesapeake Bay Program was necessary.

After the EPA released its technical synthesis in 1982, the governors of Virginia and Maryland, Charles Robb and Harry Hughes, with support from the bistate Chesapeake Bay Commission, agreed to host another conference on the Chesapeake Bay, which would eventually be called "Choices for the Chesapeake." The conference would culminate in a ceremony for the signing of the first Chesapeake Bay Agreement. Maryland and Virginia had begun to cooperate more effectively since the creation of the Chesapeake Bay Commission, but the EPA's research underscored the significance of nonpoint source agricultural pollution from Pennsylvania, which meant that without

Pennsylvania as a willing partner there would be no chance of reversing the damage done to the Chesapeake. With CBF and its congressional allies working at the federal level to ensure that the EPA published its final reports on time and remained committed to being a partner in Chesapeake environmental policy, it was left to Governors Robb and Hughes to bring Pennsylvania into the fold. In June 1983 the chair of the Choices for the Chesapeake Conference reported to the planning committee that "active negotiations at the gubernatorial level are underway to secure Pennsylvania co-sponsorship."[25] Not only were Governors Robb and Hughes asking Pennsylvania to potentially spend taxpayer dollars to benefit their two states but they were asking Pennsylvania's Republican governor, Dick Thornburgh, to risk crossing the leader of his party.

Whatever magic words Governors Robb and Hughes said to Thornburgh remain unknown, but there were compelling reasons for Thornburgh to bring Pennsylvania into the mix. For one, many Pennsylvanians genuinely cared about the Chesapeake Bay. It is a short drive from Philadelphia or Harrisburg down to the Bay, and a tour of restaurants in south-central Pennsylvania would reveal no shortage of Maryland blue crab on the menu or french fries generously dusted with Old Bay seasoning. Beyond recreational and culinary interests, the backlash to Gorsuch's handling of the EPA in general and the Chesapeake research program in particular created widespread regional awareness of all things Chesapeake. The cause attracted attention from national luminaries like the famous newsman Walter Cronkite, who moderated a pair of public forums on the Chesapeake Bay. In addition to overtures from Maryland's and Virginia's governors, state legislators from both states reached out to their counterparts in Pennsylvania about turning the bistate Chesapeake Bay Commission into a tristate organization. Ann Swanson, who worked for CBF before becoming the longest-serving executive director of the Bay Commission, recalled that two Pennsylvania legislators, John R. Showers and Kenneth J. Cole, both Democrats representing the central part of the state, were instrumental in bringing Pennsylvania into the commission and into the broader regional project to save the Bay. As Swanson remembers,

> They got involved well before the signing of the Agreement. They started coming down to the Chesapeake Bay Commission meetings, just on their own, to see what it was all about and to determine if they thought their

state should be a part of it. And the members of the Chesapeake Bay Commission were very, very open to them and inclusive of them, inviting them to dinner and to join them for their meetings. They were impressed and over time, Representatives Ken Cole and John Showers joined forces with ever-powerful Senator Dick Tilghman, who was a [Pennsylvania state] Senator with deep family roots on Maryland's Eastern Shore—think Tilghman Creek on the Chester.... Eventually they pushed to amend the Chesapeake Bay Commission legislation to include Pennsylvania, securing its passage in Pennsylvania as well.

Although it would take until 1985 for Pennsylvania to officially join the CBC, the momentum to do so was well in place by 1983. With bipartisan support across both houses of the Pennsylvania state legislature for closer cooperation with Maryland and Virginia, with Ronald Reagan withdrawing his opposition and nominating a serious EPA chief committed to the Chesapeake, and with the lure of potential millions of dollars in federal support for Bay cleanup efforts, it was only a matter of time. On June 20, 1983, Governor Thornburgh held a Chesapeake Bay Summit with his counterparts to the south. He announced that Pennsylvania would be a full partner in the upcoming conference and recovery efforts because the Chesapeake Bay was "a natural treasure of extreme importance to this and future generations."[26]

The combined efforts of state, local, and national figures to create a mechanism to act on behalf of the Bay culminated in the Choices for the Chesapeake Conference, hosted by George Mason University on December 7–9, 1983. The keynote speakers were the legendary oceanographer and filmmaker Jacques Cousteau and the Pulitzer Prize–winning author of *Beautiful Swimmers*, William Warner. Many accounts note that the conference was historic for bringing together the three watershed states for the first time and for the signing of the Chesapeake Bay Agreement. For example, the official postconference report reads, "The crown jewel of the conference was the agreement to create a Chesapeake Executive Council that will continue to coordinate regional management efforts in the future." This executive council, one of the components of the Chesapeake Bay Agreement, was certainly important, and the gathering of the states was indeed historic, but most accounts miss what really set this event apart from previous efforts to coordinate Bay policy: the genuine, overwhelming public support and interest in the minutiae of

Signing the Chesapeake Bay Agreement, 1983. *Seated from left,* Governor Charles S. Robb, Commonwealth of Virginia; Governor Harry Hughes, state of Maryland; Lieutenant Governor William Scranton III, Commonwealth of Pennsylvania. (Courtesy of Chesapeake Bay Program)

Chesapeake issues. Jacques Cousteau aside, the conference agenda was fairly straightforward. Government dignitaries gave speeches, working groups reported on their findings, and attendees participated in discussion groups to make recommendations to policymakers. This was basically what had happened at previous Chesapeake conferences, which had been somewhat sleepy affairs. The Choices for the Chesapeake Conference was like a sold-out rock concert by comparison. A *Washington Post* editorial called the conference "a stunning success, with crowds of more than 700," despite being skeptical of the ultimate intentions of the politicians. The Newport News, Virginia, *Daily Press* painted a more vivid picture: "Press people crawled all over each other. Newspaper and TV cameramen lugged their gear; some reporters held out recorders and microphones, others scribbled furiously into their notebooks. The frenzied journalists made the event look like a treaty signing; it lacked only the obligatory toasts to friendship and mutual purpose." Conference organizers had to turn people away at the door. According to Swanson, "It was just unbelievable. This gigantic room was vibrating with excitement."[27]

We now know that the conference was a major turning point and that the signing of the Chesapeake Bay Agreement inaugurated a new era of cooperation in Bay policy. At the time, however, the conference was greeted with skepticism in some quarters, especially by those who had paid close attention to Bay affairs. The *Post* editorial was skeptical that politicians would back up the big talk with big bucks. "So pardon us if we harbor some doubt that tangible results will flow from what on other counts has been an impressive summit/symposium on the bay," the paper editorialized. Likewise, the *Daily Press* agreed that "the 'treaty' will mean little if participants don't settle the funding matter." The *Daily Press* believed that Bay supporters "must sell the bay program to the president, or the states themselves must summon their political will, get the dollars, and do the job." The question of who will pay for the Bay's recovery, as articulated by the *Daily Press*, has been the greatest challenge of the Chesapeake Bay environmental movement since 1983. Although the early funding was a small fraction of what would come later—which is still short of what Bay activists say is needed—the major shock after the conference was that all the signatories contributed financially, setting an important precedent for the years to come.[28]

In the aftermath of the conference, Chesapeake environmentalists could only reasonably count on Maryland under Harry Hughes, a longtime environmental leader, to come through with significant dollars. Whether Virginia and Pennsylvania, let alone the federal government, would commit to financing the Bay's recovery was unclear. The initial response from the federal government was the most disappointing. William Ruckelshaus, despite being an ally, reported at the conference that the Reagan administration's position was that "ultimately, it is the citizens of these states—the major beneficiaries of a healthy bay—who must be prepared to assume primary responsibility for protecting their own interests." When pressed, Ruckelshaus joked that he had come to the conference "without his checkbook." All the EPA could commit to at the time was maintaining their office in Annapolis to help with monitoring and public participation. This, to be fair, was more than Bibko or Gorsuch had been willing to commit to, but it was an underwhelming offer. However, Ruckelshaus hinted that he was optimistic that he could get more support from the Reagan administration if the states led the way. Ruckelshaus was making a calculated move not only to encourage the states to follow through on their commitments but also to nudge senators and members of Congress in the Bay states to support larger appropriations for the EPA.[29]

As expected, Maryland came through in a big way for the Chesapeake with a package of laws appropriating more than $40 million, including a highly controversial law to restrict shoreline development. Pennsylvania made a symbolic commitment of $1 million to combat agricultural pollution. The biggest breakthrough, however, came from Richmond. Virginia's example highlights just how much the political tides had reversed in a relatively short amount of time. At the conference, Governor Charles Robb had proposed a $6 million outlay for the Chesapeake. William Baker had been unimpressed by Robb's proposal and publicly called him out. Saying that the figure was "very low," Baker told the press that "Robb has underestimated the willingness of the people of Virginia to pay for the cleanup of the Chesapeake Bay." Years later Baker would recall with a chuckle that after he went after Robb, "all hell broke loose and I had a target on my back in Virginia." At almost any previous time, a $6 million proposal from Virginia would have been a cause for celebration, but Baker understood that public support for the Bay was at an all-time high. Virginia's citizens would prove him right, and the state's General Assembly ended its 1984 legislative session by appropriating $13 million, more than twice what Robb had requested, for protecting and restoring the Chesapeake Bay.[30]

With the states making their financial commitments, the major question was whether the federal government would step up for the Bay. Despite appointing Ruckelshaus to replace Gorsuch and stop the bleeding at the EPA, Reagan remained opposed to funding significant involvement in the Chesapeake's recovery. Three key factors changed his mind. First, by all accounts William Ruckelshaus took a personal interest in the Bay, which included slapping a "Save the Bay" bumper sticker on his car. One Republican aide said that "Ruckelshaus made the decision to go to bat for the Bay." Second, the commitments by the states after the conference to spend millions of their own taxpayers' dollars on the Chesapeake's recovery allowed Ruckelshaus's arguments to carry the day inside the administration. It was an opportunity, according to Charles Mathias, "for the administration to demonstrate its awareness of the environment, and at the same time, show how its New Federalism program has been a working success." With Mathias, Ruckelshaus, and others members of his party, such as Senator John Warner, who was running for reelection in Virginia, advocating for the president to make the Bay a signature issue, Reagan finally relented. Gently eliding the fact that he had supported slashing its budget in the first place, Reagan told Americans in his 1984 State of the Union, "I have requested for EPA one of the largest percentage budget increases of any agency," and as we have seen, he called the Chesapeake Bay a special national resource and promised his personal support for its recovery.[31]

Yet, questions remained as to whether Reagan would fully back the Chesapeake Bay Program, especially after the pressure of a reelection campaign had ended. The third key factor, the watermen of the Bay, was perhaps the most impactful for Reagan. On July 10, 1984, Reagan toured Maryland's Eastern Shore, visiting the majestic Blackwater National Wildlife Refuge, where he met a baby bald eagle, and inspecting the docks at Tilghman Island (see map 2), where he heard from watermen about their experiences. Although much of the press coverage treated Reagan's visit as a cynical photo op, what might be called "greenwashing" today, in White House video footage from the tour Reagan appears to be genuinely moved by the difficulties facing the watermen. Indeed, these fishers must have fit Reagan's stereotype of the rugged individual, earning their living with their hands, their hard work, and their skill and knowledge of the Bay.

In one exchange, a watermen named Bill Cummings begins by bluntly telling the president, "What we're worried about is pollution on the Bay . . . when I was a kid we never had any of this at all." Cummings explained the mounting difficulties, ending with a request that was all the more powerful for being so understated: "Anything you can help us on with the clean water bill on the bay, we'd certainly appreciate it." Reagan, who had been listening intently, turned away somewhat sheepishly, gestured to EPA chief Ruckelshaus, and told Cummings, "We're going do something about that." There is a risk of reading too much into this interaction, but at the same time, it is plausible that meeting the watermen humanized the plight of the Bay and the people who depended on it. Larry Simns, who was present that day, was convinced that Reagan "listened to the watermen. He worked hard to understand our pain, fears, and our dreams by listening to our stories." While it is tempting to discount this incident as an example of a savvy politician working his crowd, it would be a mistake to dismiss entirely the potential impact on Reagan. Reagan's encounter with the watermen certainly did not turn him into an environmentalist, but it seems to have ensured that there would be no more threats to the Chesapeake Bay Program from his administration. In a little over a year the confluence of local, regional, and national events had combined to create a watershed moment in the Chesapeake Bay's environmental history. Reagan had indeed taken care of the Chesapeake's environmentalists, though certainly not in the way many would have expected.[32]

CONCLUSION

This chapter began with an angry letter criticizing the Chesapeake Bay Foundation at a moment of great uncertainty, and it ends with another letter doing the exact opposite at a moment of triumph. "As major Chesapeake Bay restoration programs begin, I am pleased to voice my support for the splendid work of the Chesapeake Bay Foundation," wrote Ronald Reagan to William Baker shortly before his trip to Tilghman Island in July 1984. Reagan found a way to fit CBF into his ideology, praising the group for demonstrating "the value of private sector involvement" in protecting the Bay. However, Reagan also made the following striking remark: "The Chesapeake Bay is a national treasure worth preserving for its own sake." He added, "For that reason, the strides your organization has made in providing field instruction in estuarine ecology deserve special commendation. There is no better way to foster a sense of appreciation and concern for the Chesapeake among future generations than through a well organized education program for our youth." Reagan's remarks reflected a dramatic shift over the course of his first term in office. Although the degree of his sincerity is debatable, it is still a remarkable statement of the intrinsic, ecological worth of the Chesapeake from a president known for his hostility and indifference to environmental issues. As much as anything else, the statement underscores that the Chesapeake's environmental movement had arrived.[33]

This chapter has only begun to scratch the surface of Reagan's environmental legacy and the shifts within the national environmental movement and U.S. politics in the 1980s, but the preliminary conclusions warrant further research. Most significant among the conclusions from this episode in the Chesapeake's environmental history is the fact that the so-called Reagan

Revolution of the 1980s was not quite ripe. Reagan faced opposition from many in his own party—Senators Mathias and Warner, Governor Thornburgh, even his own appointee, William Ruckelshaus—who advocated for the Chesapeake to varying degrees. This, combined with a widespread national backlash to the policies of James Watt and Anne Gorsuch, forced Reagan to adopt the Chesapeake as his environmental cause. This did not mean that environmental protection became a top priority for him, but it did show that he was a savvy enough politician to realize which way the political winds were blowing. In fact, the failure of Reagan's administration to destroy the EPA and roll back government regulations would only harden conservatives' opposition to environmental protection. Although Watt and Gorsuch had exited the historical stage, their intellectual heirs would be back, and they would be wiser because of the mistakes their forebears had made. Although the backlash was brewing, during the 1980s support for environmental protection was broadly bipartisan. There remains a great need to dig deeper into the debates about the role of the government, both state and federal, during that decade. The 1960s and 1970s have understandably been the focus of research on environmental politics, but as this all too brief analysis suggests, it is time for the 1980s to get a turn in the scholarly spotlight.

Another major conclusion from this episode is that it reinforces the book's central claim that environmental politics are not monolithic and are deeply connected to local contexts. What happens at the federal level certainly mattered in this case, because at the end of the day the federal government was the only institution capable of enforcing laws across the six-state watershed and the greatest potential source of funding for recovery and sustainability activities. However, the states clearly led, and the federal government followed. Working in tandem with allies in Congress like Senator Mathias, local organizers like William Baker from CBF, Fran Flanigan at the Alliance for the Chesapeake Bay, and Larry Simns from the Maryland Watermen's Association helped sync those efforts with grassroots and state level support. The cooperation between the governors of Maryland and Virginia, Harry Hughes and Charles Robb, may have been the single most essential ingredient in bringing the Chesapeake Bay Agreement together. However, their cooperation was a result of more than a decade of activism and scientific work after Tropical Storm Agnes devastated the region. While the national environmental movement might look to 1962 (Rachel Carson's *Silent Spring*) or 1970 (the first Earth

Day) as momentous turning points, for Chesapeake environmentalists 1983 was the "Year of the Bay." The timelines, power dynamics, and critical actors of local and regional groups sometimes look different from those of national environmental groups. As we seek to better understand the response to environmental challenges in the past and confront those of the present, we would do well to remember that public policy, like history, is driven as much by ordinary people and local concerns—what scholars call history from below—as it is by national concerns and leaders at the top of the political system. Treating them as somehow separate will obscure more than it reveals.[34]

Finally, for the Chesapeake itself, what mattered from this sequence of events was that people from ordinary folks like Bill Cummings to great politicians like Charles Mathias helped snatch victory from the jaws of defeat, flip an environmental opponent into an unlikely ally, and secure a commitment from Pennsylvania and the U.S. government to work with Maryland, Virginia, and the District of Columbia to "fully address the extent, complexity, and sources of pollutants entering the Bay." That commitment, though still not fully realized, was made by political leaders to their people in 1983. Since then, the environmental movement in the Chesapeake region has had its share of victories and setbacks, but its advocates have always been able to point to this document and its successors to demand that their officials keep their word. The Bay Agreement has been criticized for not being an ironclad law, for lacking specific goals, timelines, and enforcement mechanisms, and relying on voluntary cooperation to implement change. Part of the argument in this chapter, and indeed in this book, has been to restore the luster and significance of this promise. From the perspective of 1983, after decades of work to reach such a point, after seeing the EPA's Chesapeake Bay Program nearly killed off, to see the Chesapeake Bay Agreement transform the EPA's Chesapeake Bay Program from a one-off research study into an ongoing partnership between the EPA and states in the watershed is to see it for what it truly is: a triumph that created the possibility of coordinated action to restore the Chesapeake Bay. The fact that the remaining states in the watershed—Delaware, New York, and West Virginia—did not sign any of the subsequent agreements until 2014 and still have not joined the Chesapeake Bay Commission points to how significant an achievement it was to bring Pennsylvania into the fold in 1983. For all its flaws, the Chesapeake Bay Agreement very nearly did not come to be, and under a different set of circumstances

Maryland and Virginia might have soldiered on alone or returned to squabbling between themselves while the Bay slowly died.[35]

The next chapter looks at a few high-profile battles to implement policy changes under this new institutional framework in order to better understand its strengths and weaknesses, to observe how the movement grew, and, at long last, to see the Chesapeake begin to recover. As will become clear, again and again activists returned to the promise made in the original Chesapeake Bay Agreement to demand more. In many ways, the agreement's weakness became its greatest strength; by requiring constant vigilance on the part of the Chesapeake's defenders, it ensured that a vigorous and dedicated environmental movement would continue to grow throughout the watershed. Through resulting policy statements, the Bay Agreement crystallized a group of organizations, including the Chesapeake Bay Foundation, the Chesapeake Bay Commission, and the EPA, along with major research labs and of course the Bay states, around a set of common goals. It became the connective tissue that would link issues across levels of government and throughout the watershed. Precisely because the agreement lacked deadlines or enforcement mechanisms, the institutional framework it created allowed for cooperation, trust-building, and precedents that ultimately led to specific, measurable goals and enforcement through a consensus of political will. This wasn't necessarily a conscious goal of the signatories, but the product in 1983 reflected the process that created it. Since the 1980s the people of the region have repeatedly demanded that their elected officials save the Bay, and while progress has been at times halting, the estuary has slowly, tentatively, begun to recover. Or put differently, despite four decades of population growth, suburban sprawl, industrial agriculture, and commercial fishing, the framework created in 1983 has sustained the Chesapeake as a national treasure.

FOUR

PROGRESS AND BACKLASH

The mid-1980s marked an important point of departure for Chesapeake environmentalism. For nearly two decades, from the early 1960s to 1983, there was one overriding focus for the movement: to persuade the general public and political officials that there were grave problems facing the Chesapeake Bay and to forge a commitment to implement a coordinated, watershed-wide program involving multiple states and the federal government to address those problems. After 1983, the scope of Chesapeake environmentalism expanded dramatically. In the decade following the Chesapeake Bay Agreement, 1984–94, the Chesapeake Bay Foundation's membership grew from approximately 23,000 to 87,000, and its revenues leapt from $1.5 million to $6.5 million. While many more individual and institutional players made major contributions to the Chesapeake restoration effort, CBF's growth is a useful heuristic for understanding the broader movement as the "ecosystem" of Chesapeake environmentalism grew and diversified. At the same time, the opponents of Chesapeake environmentalism grew in number and power. Call it the sociopolitical equivalent of Newton's Third Law: the action and reaction of social forces produced gains and setbacks that were the governing laws of motion for the Bay's recovery.[1]

Rather than trying to follow one narrative thread through the heady decade following the first Chesapeake Bay Agreement, this chapter examines four case studies in detail. These case studies explore the tensions between environmentalists, commercial fishing, and resource managers to save the striped bass fishery; the failure of these same groups several years later to avert a catastrophe in the oyster industry; the pitched battle between industry-funded lobbyists and grassroots citizen activists over banning phosphates in laundry detergents; and finally the difficulty in addressing nutrient pollution from agriculture in Pennsylvania. Analyzing individual case studies allows a closer look at the mechanics of the Chesapeake's environmental movement to better understand what it looked like on the ground and in the water. It also allows a detailed look at the emerging opposition to Chesapeake environmentalism, which reflected national trends against environmental regulations. These case studies show the strength of the movement and the broad public support for the Chesapeake, but they also illustrate how easily a well-organized group of opponents can delay progress, defeat bills, and derail implementation of policy. Collectively, they show the potential and the limits of Chesapeake environmentalism, with the latter looming larger and larger into the 1990s.

Although this period ended with the full promise of the Bay's recovery unfulfilled, it was a vital period in the Chesapeake's environmental history because for the first time the great estuary's decline was halted, and in some areas dramatically reversed. In 1983 the authors of the EPA's final report, *Chesapeake Bay: A Framework for Action*, wrote that their "findings clearly indicate that the Bay is an ecosystem in decline." A decade later, the Chesapeake Bay Program's biennial monitoring report found that "vital signs, such as living resources, habitat, and water quality, are stabilized." As the authors put it, the Bay was "out of intensive care" and "on the road to recovery." There is a tremendous gulf between a "stabilized" Chesapeake Bay and a "saved" Chesapeake Bay, which explains much of the frustration that activists, scientists, politicians, and the general public felt at not being further along that road to recovery by the mid-1990s. However, sometime during the 1980s, somewhere between "stabilized" and "saved," a sustainable Chesapeake emerged. The story of Maryland's official state fish, the Atlantic striped bass, is a good place to begin to understand this new era in Chesapeake environmental history.[2]

SAVING THE STRIPED BASS

Locals call it "rockfish," which is an appropriate common name for the species, given that the latter half of its scientific name, *Morone saxatilis*, means "rock dweller." The nickname is well earned in the Chesapeake, for anglers will seek out oyster "rocks" (reefs) and other underwater obstacles to find the biggest, feistiest striped bass. As a large, anadromous predatory fish (the Chesapeake weight record is more than sixty-seven pounds, and some specimens have been up to five feet long) the striped bass is a key component of the health of the Chesapeake and Atlantic ecosystems. It is also "the watermen's number-one fishery and the recreational fishermen's number-one trophy fish," according to the Maryland Watermen's Association's Larry Simns. Some fans of the fish have gone even further, elevating the striped bass to "the aquatic equivalent of the American bald eagle." Like the national emblem, striped bass had their own encounter with population collapse and triumphant recovery. With harvests of oysters and shad declining during the twentieth century, striped bass became increasingly important for Chesapeake watermen. In 1973 Maryland's commercial fishery landed 14.7 million pounds of striped bass; a decade later the catch had decreased by 90 percent. As catches of striped bass declined, their prices went up, and despite their decreasing numbers, they became all the more essential to the economic survival of the Chesapeake's small-scale commercial fishers. "By 1983," according to Simns, "the commercial fishermen and the recreational fishermen were at complete odds with one another. I still recall how ugly and bitter the infighting was between us." William Goldsborough, who was CBF's senior fisheries scientist at the time, remembers meetings that were "as close to pitched battle as you could be without actually having one." The title of a popular book on the subject, *Striper Wars,* by Dick Russell, succinctly captures the tension and the stakes for the species, the ecosystems, and the livelihoods that depended on its survival. Although adult striped bass range across the Atlantic Seaboard from Florida to Nova Scotia, 70–90 percent of striped bass spawned in the Chesapeake Bay, making the estuary the biggest and most important front in the "striper wars."[3]

Perhaps the most important key to understanding the "striper wars" is a humble scientific survey. Every year since 1954 the Maryland Department

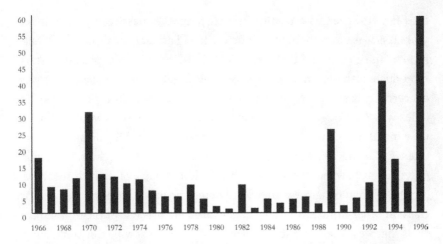

Maryland striped bass juvenile index, 1966–1996. (*Source:* Maryland Department of Natural Resources)

of Natural Resources (DNR) has conducted a survey of juvenile striped bass. Three times each summer at twenty-two fixed locations across the striped bass's major spawning and nursery areas the DNR deploys seine nets from the shore, then pulls them in by hand and counts the catch manually. At the end of the summer, the results from these surveys are compiled to produce the juvenile index. This is the highest-quality, longest-running survey of striped bass populations, and given that a supermajority of the species spawns in the Bay, the juvenile index is an essential indicator of the entire population's health. What the surveys revealed by the mid-1970s was that the spawning success of striped bass fluctuated cyclically, with a major "dominant year class" occurring roughly every half decade or so. The problem was that striped bass in Maryland and Virginia reached legal size—12 inches during the 1970s—after about two or three years but did not reach full sexual maturity until about six or seven years of age. Simply put, by the early 1970s striped bass were seriously overfished and unable to replenish their numbers through reproduction.[4]

The Chesapeake Bay Foundation first took an interest in striped bass in 1975, hosting a panel of experts "to discuss the problems and promise of sport fishing on the Bay." At the time, striped bass were not yet a major issue; in 1970 the juvenile index was 30.5, making that year's cohort the largest dominant year class yet recorded, albeit only one of sixteen classes recorded thus far. However, as a result of overfishing, the massive 1970 dominant year class was virtually wiped out before it had a chance to reproduce. Thus, in 1976 and 1977, when that large class would have been expected to start reproducing, the juvenile index was a paltry 4.9. Although that index was concerning, it wasn't the lowest there had been. Then, in 1978 the juvenile index was 8.5. Not great, but perhaps a sign that the species would be on an upswing. However, as CBF's Goldsborough recollects, fishing pressure was so intense that this cohort had no chance to reproduce. "Since we were fishing on 12 inch fish, when they hit 12 inches within a couple of years, by like, '80, '81, the population surveys they were doing showed no blip for the '78 year class. In other words, they caught them all. That was one of the first indications that we do have the ability to overfish and to just wipe this thing out."[5]

The outlook for striped bass looked bleak in the early 1980s, which led to some of the first serious bits of rancor in the press. In 1981, lobbying by Larry Simns and the Maryland Watermen's Association helped to defeat a bill that would have raised the legal limit on striped bass to 14 inches. In response, the *Baltimore Sun* published a scathing editorial that accused the MWA of playing "smart politics" but "dumb economics." The editors personally attacked Simns for putting on an act of being "so folksy as to be almost charismatic" and warned that "the watermen may find themselves in the position of the ranchers out West who agitated for the right to overgraze, won it, and as a result watched their land blow away in the Dust Bowl." In conclusion, the editors implored "the watermen to stop being smart and start being smarter." Unsurprisingly, this did not go over well with Simns and the MWA. However, what was a bit surprising is Simns shot back a defense in the pages of the *CBF News*. Writing a public letter to the editor, Simns defended himself as only a "part-time lobbyist" but "a full-time waterman making my living by oystering, crabbing, clamming, eeling, and fishing." Simns's objections to raising the legal size of striped bass pointed to a classic problem of the commons. "If the state of Maryland could put a fence up on the MD/VA line so the fish would stay in Maryland waters, then their projections would be right and there would be larger fish to catch in another year." Simns continued, "the only thing that the 14 inch limit would do is to preserve the stock in such a way so that Maryland would lose a large percentage of the catch and increase the catch for Virginia and the Carolinas." Further, Simns argued these restrictions were not based in science, but coming from recreational anglers "who will not be satisfied with anything the Department of Natural Resources does short of completely eliminating the commercial fishermen from catching striped bass in the Chesapeake Bay waters." Simns's comments in 1981 verify his later recollections of how bitter the infighting between recreational and commercial fishers was, but his choice of the *CBF News* points to an uncommon role the organization would play mediating this dispute and ultimately helping to ensure the sustainability of the fishery for commercial and recreational uses.[6]

In the midst of the increasing tensions over the dwindling striped bass fishery and against the backdrop of the growing concern about the Chesapeake Bay discussed in the previous chapter, the Maryland DNR reported the 1982 juvenile index was once again 8.5. This was perhaps the last chance to preserve the species at commercially viable levels. Indeed, a 1992 study of the age of the striped bass population—the age of the fish can be determined by counting the rings in the earbones, akin to counting tree rings—did not find a single fish spawned from 1972 to 1981.[7] Not only were all the adult fish wiped out but the chances of any other significant striped bass reproduction were slim to none. The 1981 and 1983 juvenile indexes were the two lowest on record. If the species was to recover, the 1982 cohort had to reach maturity and spawn. Goldsborough recalled that "there was a momentum building to say, 'Wait a second. We should protect these.'"[8] The interest in protecting the 1982 class of striped bass extended beyond the watershed. Recognizing that this was a crucial lifeline for the species, in 1983 the Atlantic States Marine Fisheries Commission (ASMFC) proposed a plan that called for reducing the coastwide harvest by 55 percent. The commission had no regulatory authority to enforce the plan, so it was up to the individual states to voluntarily reduce their harvests; more importantly, there was no scientific evidence that a 55 percent reduction would actually restore the species. The ASMFC's proposal delegated responsibility to the states, which put Maryland and its traditional harvest of 12-inch, immature striped bass at the center of the debate over how to manage the species.

With the clock ticking as the 1982 cohort approached legal size in Maryland, commercial and recreational fishers were pointing their fingers at each other, but most stakeholders were not yet calling for an outright moratorium. For instance, in written testimony regarding updated striped bass regulations submitted to the DNR on August 21, 1984, CBF stressed that "it is especially important to protect the existing stock of 1982 year-class fish . . . this stock represents our best hope for natural reproduction in the years to come. It must be preserved." However, the authors added, "we do not feel that we can recommend a complete moratorium on fishing for striped bass . . . therefore, at this time, we reluctantly support the effort to reduce striped bass harvest by 55% throughout the Atlantic coastal states." Instead of a total ban, the foundation suggested harsher penalties for violators, including taking away the licenses of repeat offenders. Less than a month later, Torrey Brown, the DNR's secretary, made a decision that "surprised virtually everyone interested in how to best manage striped bass." On September 11, 1984, he announced a statewide ban on the entire striped bass fishery—recreational and commercial—beginning January 1, 1985.[9]

Brown's decision was stunning because it symbolized the collective failure to save the fishery. Brown officially declared striped bass a "threatened" species, which precluded any harvesting whatsoever. CBF's newsletter called it "a sad day," and Goldsborough explained that many at CBF saw it not as a victory but as a tragedy because the vitality of the Chesapeake's unique fishing communities was "one of the most tangible examples of the Bay and why we care about the Bay and why we're trying to save the Bay." Larry Simns called it "the lowest point of my professional life." Many watermen were enraged. On December 31, 1984, the last day the fish could be legally harvested in Maryland, watermen filled a pickup truck with dead striped bass and left it parked in front of the Maryland State House. Why did Brown admit defeat and close a fishery that nearly all parties involved wanted to find a way to keep open? The juvenile index. In early September Brown got the survey results, an anemic 4.2. There were no reinforcements coming. The 1982 cohort would make or break the striped bass.[10]

Results from Maryland's juvenile index also undercut a central premise of the ASMFC's plan. The commission assumed "a doubling of egg deposition within ten years if a 55 percent coastal harvest reduction were implemented." The problem was twofold. Maryland's juvenile index indicated that fish survived at a much lower rate than the ASMFC's model assumed, and there was no evidence whatsoever for a linear model of striped bass reproduction. The only surefire way to rebuild fish stocks was to protect the 1982 year class long enough for it to reproduce in the right environmental conditions for another dominant year class—at least six years in a best-case scenario. Even earlier proposals, like the one CBF supported, raising the legal size in Maryland from 12 to 14 inches, would not have protected enough fish long enough for the 1982 year class to survive to full reproductive maturity. Other states, especially Virginia, contested Maryland's index. The Virginia Institute of Marine Science objected to Maryland's findings because VIMS data did not show a decline in recruitment numbers but rather low but stable recruitment. The problem was that VIMS's data were not as reliable as Maryland's. The VIMS survey was similar to the Maryland survey, but it began in 1967, was cancelled in 1973, and then restarted in 1980. In other words, VIMS did not have the baseline data from the 1950s and 1960s, and Virginia, along with many other East Coast states, inclined to believe the ASMFC's model, did not plan to follow Maryland's lead and close its fishery. Thus, interstate rivalry nearly doomed one of the Chesapeake's iconic species.[11]

Maryland's ban shifted the battlefield to the U.S. Congress. In October 1984, a month after Torrey Brown announced the ban, the ASMFC held its annual meeting. By one estimate the individual states' plans would only amount to "a 25 to 35 percent reduction in landings." Voluntary measures would not work, and Maryland's leadership, together with the other states' inability to come up with a semiplausible plan, led Congress to act. The House of Representatives passed a bill introduced by Gerry Studds (D-MA) that would require states to follow the ASMFC's striped bass management plan or face a federally imposed moratorium. After some Senate wrangling, led by John Chafee (R-RI) and Ted Kennedy (D-MA), and another House vote, the Atlantic Striped Bass Conservation Act passed Congress on October 31, 1984. It was a shockingly quick turnaround by congressional standards, and in an election year no less. Ultimately, Maryland's ban was the reason for such swift action. It was especially significant because it would protect the striped bass during the spawning run, when they were at their most vulnerable. Maryland's self-imposed total ban made it that much easier for Congress to force the other states to abide by a 55 percent reduction in the harvest. Although it is a rather obscure law now, at the time of its passage it was groundbreaking. "For the first time ever," to help save striped bass, "Congress had stepped into a fisheries management crisis in traditional state waters."[12]

Although CBF initially did not call for the harvesting ban, once it was in place the foundation was a vocal supporter for keeping it in place and extending it. CBF's most critical role, however, was as a mediator between the recreational and commercial fishers. The ASMFC's strategy to protect the 1982 class until it reached maturity meant that in practice most states chose to raise their legal size limits to stay ahead of its growth, while not officially banning the harvest of striped bass. Virginia, for instance, raised its legal limit by several inches each year before finally closing its fishery in 1989. New England states such as Massachusetts and Rhode Island, which traditionally harvested adult fish (in contrast to the Chesapeake's heavy reliance on what might playfully be called "teenage" fish), kept their fisheries open but raised legal sizes to as high as 38 inches by 1990. These measures successfully reduced harvest pressure on the 1982 cohort, and in a triumph of biology and bureaucracy, the 1989 juvenile index was 25.2, the second-highest tally recorded in the thirty-five-year history of Maryland's survey. In 1993 the juvenile index set a new record of 39.8, which was subsequently shattered in 1996 with an index of 59.4, the current record.[13]

Reducing harvests overall and protecting the 1982 class had kept enough of those fish alive to spawn and produce a dominant year class with the right environmental conditions. (The subsequent juvenile index in 1990 would be a bafflingly low 2.1, underscoring how much had to go right in 1989.) But protecting the 1982 cohort was only half the battle. Ensuring that the fishery could reopen and be managed sustainably was a daunting challenge, given that the acrimony between commercial and recreational fishers remained at a boiling point. Goldsborough said that because of the dynamic between the different groups of fishers, many people worried that "this is going to be a dogfight, this is going to be a bloodbath, we're not going to really be able to make progress this next step when we reopen [the fishery]."[14] Consequently, he began arranging informal meetings between sport fishers, charter boat captains, and commercial fishers to come up with a joint strategy for reopening and sharing the striped bass fishery.

Largely through Goldsborough's personal diplomacy, CBF acted as a mediator between competing resource users, a role environmentalists rarely played in other fishery crises. For instance, as Joseph Taylor has shown regarding the Pacific Northwest's salmon fishery, "Environmentalists don the garb of *Most Worthy Protector* in order to speak for salmon and The People" through a "demonization of rivals" that casts opponents as "unambiguous villains."[15] Often, environmentalists formed alliances with sport fishers; not only were sport fishers often of the same white, middle-class background as environmentalists but in fact many recreational anglers founded and supported environmental groups in the 1960s. In the Pacific Northwest, indigenous rights to harvest salmon added a wrinkle missing from the drama over striped bass, but CBF's role in the striped bass debates is still instructive as an alternative to the roles environmentalists typically play in these contests.

As Goldsborough tells it, he went to the major associations—the Maryland Watermen's Association, the Maryland Charterboat Association, and the Maryland Saltwater Sportfishermen's Association—and he would "get one or two guys from each of them, and we met in the upstairs back room of this bar down at city dock ... what we did several times, just trying to sort of create a rapport, break down barriers, talk conceptually about where we might go." Goldsborough's informal meetings laid the groundwork for trust between the groups in a more formal institutional setting. In 1990, Maryland's DNR created a Striped Bass Advisory Board, with Goldsborough as chair, to help create fishing regulations with input from the main fishing groups. Goldsborough's informal work paid off in the late 1980s, as the fishers reached a tentative truce to keep the fishery closed through 1989. Maryland reopened its fishery in 1990 under regulations hammered out by the three groups. Initially, Maryland charter boat captains successfully lobbied to take five fish a day, but the harvest pressure was so great that the season was shut down in less than two weeks. In a sign of the fragility of the species (and the resiliency of the group's trust), the stakeholders agreed that "two fish a day at 18 inches was fine for maintaining a charter fishery, and a recreational fishery, and then a quota on the commercial side." At the time, the striped bass regulations reflected a remarkable balancing act between the needs of the fish and the needs of the fishers.[16]

CBF's role as a mediator became especially important when a faction of recreational anglers found sympathetic legislators to introduce a bill that would declare striped bass a game fish in Maryland and outlaw their sale, thus permanently ending the commercial fishery and confirming Larry Simns's deepest suspicions. Claiborne W. Gooch III was a member of the Maryland Saltwater Sportfishermen's Association, and he wanted CBF's support for his bill. Gooch typified many recreational anglers—after a twenty-year career with IBM, he retired to Maryland's Eastern Shore—and considered himself "a friend of the fish." Like many of the participants in the Pacific Northwest's salmon crisis, Gooch had "artfully converted self-interest into principle."

Among Gooch's many arguments why CBF should support his bill, he wrote, "There are not even enough fish to allow the 500,000 anglers of the Bay in Maryland even one fish apiece." Gooch pleaded for CBF's support because he recognized that with its status as the region's leader on environmental issues, the foundation had the power to "make or break that issue." In a private memo to CBF's board of directors, Goldsborough suggested a fine line. "It should be made clear that we do not *oppose* the proposal," he wrote "we simply do not believe CBF should be involved, and we think it is important to the integrity of our role in the fisheries not to be drawn into such a dispute." Goldsborough felt that CBF should only focus on the conservation of the fishery and not get involved in disputes about the allocation of the catch. CBF's neutrality on the issue amounted to tacit support for the commercial fishers' rights to a share of the fishery, but more importantly, it enabled the foundation to keep peace between the sides and maintain its reputation as an honest broker.[17]

Since the fishery reopened in 1990 the striped bass has been repeatedly held up as one of the major success stories of the Bay's restoration. Recent data cloud this rosy outlook, with the 2023 juvenile index the second lowest of all time, 1.0, continuing a run of bad years for the striped bass. The ASMFC, CBF, and the Chesapeake Bay Program have all reported that the striped bass are being overfished and that they are suffering from overfishing of their major prey species, menhaden. It is important that current fisheries managers not squander one of the Bay restoration movement's signature achievements. Once on the verge of commercial extinction in 1984, a decade later the Atlantic striped bass was "declared restored to historic levels." Equally important, Maryland watermen were able to participate in a commercial harvest as well. It would have been a hollow victory for Chesapeake environmentalism if the striped bass were restored at the cost of the commercial fishery.[18]

A number of interesting conclusions may be drawn from the story of the striped bass. First, as with the Chesapeake Bay Program as a whole, the impetus for federal action came from the state level. In this case, it was Maryland leading the way with a state-level ban, which allowed Congress to pass the Atlantic Striped Bass Conservation Act. Second, consistent, baseline scientific data are exceptionally valuable to fisheries managers, especially if they are willing to act on the data. Torrey Brown deserves a bust in the Chesapeake Bay's Hall of Fame for his courageous decision to shut down the fishery to save the 1982 cohort of striped bass, but equally important were all the DNR employees who year after year collected the data to support Brown's decision. Finally, a crucial lesson for environmentalists is that they don't always have to pick sides. As we will see, not every fishery was amenable to such solutions, but the striped bass's history stands as a reminder that such solutions are possible.

FAILING THE OYSTERS

The success of restoring the striped bass fishery makes the failure to halt the crash of the oyster fishery in the 1990s all the more tragic. Unlike striped bass, oysters stay put (at least adult oysters do). If Maryland or Virginia decided to protect oysters in their portions of the Chesapeake, they would not wind up being harvested by fishers from North Carolina or Massachusetts. Likewise, while striped bass need six or seven years to reach full reproductive maturity, oysters can begin reproducing in two years or less. Furthermore, oysters' role as a keystone species capable of filtering the Bay's waters and providing crucial habitats via their reefs made their recovery a central pillar of the Bay restoration plan, and arguably made the oyster even more important than the striped bass to the economic, environmental, and cultural integrity of the Chesapeake. Despite having some advantages over the striped bass and an urgent need for action, the lack of a clear scientific picture of what was happening to the oyster combined with the deadly diseases MSX and dermo to wipe out the fishery. In 1990, the same year that Maryland lifted its striped bass ban, "oyster abundance in the Chesapeake Bay [was] at its lowest level in history." Sadly, this was not yet rock bottom, as conditions got a lot worse for the Bay's beloved bivalve, with harvests bottoming out in the 2003–4 season at a scant twenty-seven thousand bushels. Although oysters are beginning to show some encouraging signs of recovery, in contrast to the story of the striped bass, which showed the potential of Chesapeake environmentalism to sustain the Bay, the failure to protect oysters stands as a glaring missed opportunity.[19]

The history of oysters in the Chesapeake Bay has already been well documented, with many excellent books on the subject, led by Christine Keiner's

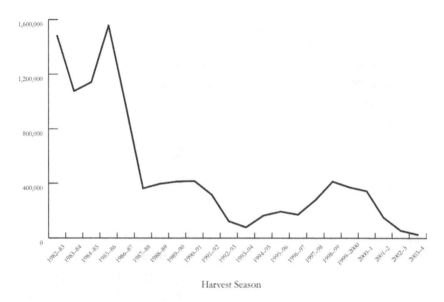

Maryland oyster landings, 1982–2004. (*Source:* Maryland Department of Natural Resources)

The Oyster Question. The oyster has tremendous significance as the Bay's original boom-and-bust fishery. At its peak in the decades after the U.S. Civil War, what was possibly the world's largest oyster fishery employed more than fifty thousand people, and annual harvests routinely topped 10 million bushels. Standout years, such as the 1891–92 season, when the Chesapeake harvest totaled 17.6 million bushels and accounted for three-fifths of the total U.S. harvest, were not uncommon. Even with oyster harvests at their height, scientists and government officials were concerned about how to best manage the resource, making oysters the Bay's "first and historically most important conservation issue," as Keiner has persuasively argued. Although oyster harvests fell to a third of their nineteenth-century highs by the middle decades of the twentieth century, Keiner has shown that for a time oyster harvests were sustainable. As she found, at least for the Maryland portion of the Bay, "scientists, watermen, and policymakers worked together to create a unique system, a regulated common that balanced science-based conservation and dignified self-employment." This system fell apart in the 1960s and 1970s as poor water quality and the arrival of MSX and dermo made harvest pressures unsustainable. As with so many things related to the Chesapeake, Tropical

Storm Agnes marked a turning point: harvests have never matched levels recorded before the 1972 deluge buried oyster reefs under millions of pounds of sediment. Thus, by the late 1980s the oyster industry was already a shell of itself. Owing to the combination of overharvesting, poor water quality, and diseases, the oyster fishery was in the same state the striped bass fishery had been just a few years earlier, being "totally dependent on year to year reproduction as oysters are caught as soon as they mature to legal size."[20]

With the recovery of the striped bass occurring as oyster populations dipped lower and lower, many people quite naturally began to ask whether what had worked for the bass would work for the bivalves. CBF staff began to hint that a moratorium might be necessary, but they were reluctant to make a policy recommendation banning oyster harvests. For instance, in 1989 CBF president William Baker wrote that "Maryland and Virginia will be forced to take drastic action soon to save the oyster industry. . . . If we do not bite the bullet now, a total moratorium will be the only way to save the oyster, an absolutely essential step in saving the Bay." Years later, William Goldsborough explained that CBF was "reluctant to call for a moratorium of any sort, other than as a last resort, because it is a blunt instrument of management, it can be destructive of some things you might be trying to preserve." The foundation wanted to preserve fishing communities as part of the social and environmental tapestry of the Bay. At the time, CBF was not yet calling for the striped bass fishery to reopen, and closing the oyster fishery at the same time as the striped bass fishery would have been fatal to the livelihoods of men like Larry Simns and communities like Rock Hall or Smith Island, which depended on commercial fishing.[21]

In June 1991, when it was clear the striped bass fishery was back and would remain open, CBF finally demanded that Maryland and Virginia implement a three-year moratorium on their oyster fisheries. CBF acknowledged that this would place "an enormous burden on the Bay's 2,000 full-time oystermen" and called for "licensed oystermen [to] be employed to participate in oyster repletion programs, research efforts, and other publicly supported measures to rebuild the fishery." Harvests had continued to decline, but the fishery was caught in an economic and biological death spiral, with diminishing returns increasing the value of each remaining oyster, driving up harvest pressure when the species could least sustain it. A top-ranking Maryland fisheries official considered the foundation's request for a moratorium "out of the question" despite a streak of record low harvests, because the industry still brought in $25 million a year. Goldsborough took to the op-ed pages of the *Washington Post* to defend the foundation's plan. Citing recent scientific evidence that the oyster populations were at 1 percent of their historic population levels, Goldsborough argued that "the famous oysters of the Chesapeake Bay are on their way to commercial and maybe even biological extinction." Oyster harvests had declined by 80 percent in Virginia and 90 percent in Maryland over the course of the 1980s, as the state's usual approaches to conservation failed to prevent a collapse. "Historically," Goldsborough wrote, "Maryland and Virginia have relied on declining profits to restrain the fishery from taking excessive numbers of oysters. But that philosophy has been undermined by the high percentage of adult oysters that are dying from disease in most of the high salinity parts of the bay. A few oysters with some natural ability to resist the diseases persist, and these are the ones that in many areas support the remaining fishery." Without protection for the few remaining oysters that had some resistance to MSX and dermo, the states' "traditional approach [to management] will lead to an end to the fishery." "A new philosophy is needed," he argued, "that says we will conserve as many of these oysters as we can so that they may serve as the basis for rebuilding the population. The surest means to this end is a temporary moratorium on oyster fishing."[22]

Unsurprisingly, one of the voices pushing back against CBF's proposal was Larry Simns of the MWA. In his own *Post* editorial, he conceded that oysters were in trouble but charged that "the foundation [CBF] did not, however, give enough weight to the real villains in the case of the bay's oyster loss—parasites, high salinity and poor water quality." "The main problem," he wrote, "is not that overharvesting has reduced the number of oysters in the bay. An annual survey conducted since 1939 has shown that the 'spatfall' (the amount of oyster larvae that set or attach themselves to clutch) continues to be large enough to sustain the traditional 1 million- to 2 million-bushel annual harvest." Simns's contention that a harvest of one to two million bushels was "traditional" was puzzling, because 1987 was the first time in one hundred years that oyster harvests had fallen below two million bushels. Simns insisted that "the problem is that not enough oysters are surviving, which has nothing to do with harvesting practices." Goldsborough recalled that despite trying to be sympathetic to the plight of "oystermen," CBF was "vilified" in fishing communities. He explained that whereas CBF had maintained a working relationship with commercial fishers over striped bass, "you didn't have recreational fishermen involved in oysters at that time"; as a result, environmentalists and watermen became "warring parties at that time with respect to oysters."[23]

Thanks in part to the journalist Tom Horton, there is a remarkable window into the state of these "warring parties" in the late 1980s. In 1988 CBF organized a panel featuring three of the most eminent Chesapeake scientists, Eugene Cronin, William Hargis, and Donald Pritchard to discuss the Chesapeake's ecological problems at a retreat for CBF's environmental education staff (see chapter 2). Horton had the foresight to record the discussion, which included an open question-and-answer session, on a VHS tape. The audience at the event included Captain Ed Farley, a commercial oyster fisher, and Nick Carter, a biologist from the Maryland DNR. When the discussion turned to oysters, the conversation got very heated very quickly, and both Farley and Carter jumped into the fray to defend their points of view. More than thirty years later, the passion and emotion these men felt for the Bay in general and oysters in particular is clear. Speaking at a time when oyster harvests in the Bay were at a then record low—little did they know how much further they'd fall—William Hargis, who had spent twenty years as director of the Virginia Institute of Marine Science, vented his frustration at the inability to rein in harvest pressure on oysters. "When you talk about overfishing," he said, "you're pointing at the watermen and they become sensitive, naturally. And they say, 'well, the environment is going downhill.' And I say, 'yes, it is. But, the most manageable element in the whole equation is fishing effort!'" His face growing redder, Hargis continued to complain about the failure of fishery managers to listen to science. Working to an emphatic conclusion, Hargis cried out, "The facts are that we have enough scientific information about oyster biology and the natural forces affecting oysters and we have enough technology . . . we could bring back oyster production in the next 5 to 7 years to pre-MSX periods. And I think to even pre-1900 periods. And I don't expect the recommendations to be followed!"[24]

After Hargis finished speaking, Ed Farley spoke up. Farley was a transplant from New England who had fallen in love with sailing the Chesapeake as a young man and worked his way up from deckhand to captain, eventually owning his own skipjack, the *Stanley Norman*. He had been invited to attend the panel discussion because he had partnered with the Chesapeake Bay Foundation to take children out on his boat to show them the traditional way of harvesting oysters under sail power and to help nurture an appreciation for the Bay's unique maritime culture. Farley was unhappy with Hargis's quick dismissal of the watermen. He challenged the exclusiveness of scientific knowledge, arguing that "the watermen have been out there monitoring it every day, with the same boat and gear, no scientific evidence, but have been totally put down by the scientific community, as not knowing anything, and it's true that they don't know how to evaluate what they're monitoring, but they are not recognized as a source of information." After a back-and-forth with Eugene Cronin about how much the scientific community respected the watermen, Farley attempted to offer an olive branch by suggesting that the problem was not how much or how little the scientific community respected the watermen but how science was conveyed by state officials. "To get some of the scientific information and ideas across to the watermen, it is usually presented by DNR, not by the scientific community," he said. "And DNR is seen as the enemy." Farley questioned the DNR's credibility, saying that "you're not going to listen to what DNR tells you because they mismanage with their decisions, and we would gladly take the good information that the scientific community has to offer."[25]

At this point, another audience member, Nick Carter, leapt to his feet, half-joking that someone had "to stand up for these villains at the DNR." Carter, as it so happened, was one of those "villains" at the DNR, but he was no ordinary fisheries biologist. Throughout his career, Carter defied the stereotype of the pencil-pushing bureaucrat by challenging his superiors and taking direct action to help the Bay. In a retrospective article, Tom Horton described Carter as someone who "unfailingly put the Chesapeake Bay's natural resources first, letting the chips fall where they might. His unflinching comments often made life tougher for a department under pressure to let development and fishing proceed as usual." A senior CBF vice president, Don Baugh, called him "a teacher, a mentor, a spiritual adviser." What frustrated Carter on that day in 1988 was not that a waterman was demonizing the DNR—he was accustomed to that—but that the scientists wouldn't clearly identify the problems facing the oyster industry. Carter cited the case of the oysters in St. Mary's River as a prime example. Speaking with increased pace and emotion, Carter said that the DNR had turned to three academic shellfish scientists for an explanation of the oysters' decline, and one had blamed it on sewage, another on pesticide runoff, and the third on overharvesting. At that point Hargis shouted that "all three would be right!" Carter shouted right back, "I know! But what it does, is it allows somebody who's got to make a hot decision to sit on the fence!" Feeling the need to defend scientists' role in policy formation, Cronin jumped into the fray, arguing that "as long as you have insufficient information, limited information, you can take any of several positions, and each is equally consistent . . . none is more right than the others." Carter was unfazed. "What you have to understand," he pleaded, "from a management angle, if you go against somebody you've got to be ready to prove your case to the hilt." With Cronin, Hargis, and Carter practically at one another's throats, the third panelist, Donald Pritchard, tried to defuse the tension with broad, periphrastic comments about the nature of scientific inquiry. Shortly thereafter Horton wisely called for a ten-minute recess to allow tempers to cool.[26]

What this episode shows is that in the case of oysters, unlike in the case of the striped bass, there was nothing as definitive as the juvenile index that could prod politicians to action. This debate also highlights the building tensions around the oyster fishery, which would only worsen into the early 1990s. In the 1970s, watermen like Ed Farley or Larry Simns could champion an EPA study out of concern for the Bay and out of self-interest; recall that from their point of view, a report that pointed to water quality took the heat off commercial fishers. Into the mid-1980s, watermen could see the value of environmentalists as a counterweight to sport fishers, but by the late 1980s and early 1990s the progress toward managing the Bay's fisheries increasingly set the two groups against each other, which ultimately led to failure. Even after the Virginia Institute of Marine Science publicly sided with CBF in 1992 and recommended that the Virginia Marine Resources Commission institute a moratorium on oyster harvesting, William Pruitt, the commissioner, insisted, "We're in good shape for this year." Pruitt was wrong, and 1992 set the oyster bar even lower as the lethal combination of pollution, overharvesting, and disease decimated oyster populations. In 1993, Virginia's Marine Resources Commission at least held a hearing to consider a moratorium—it was out of the question in Maryland—but "in spite of strong scientific and public support, the commission rejected its own staff's recommendation to close the public fishery for market-sized oysters." The confrontation with the watermen over closing the oyster fishery was, in William Goldsborough's words, "highly traumatic and unfortunate," and in the end neither Maryland nor Virginia chose to follow the striped bass playbook. It would be another decade of decline before oyster harvests hit rock bottom, and new, costlier approaches to reviving oyster populations would be required.[27]

More than thirty years after the Bay states failed to take strong actions to save the oysters, there are reasons to hope that aquaculture, restoration, and sanctuaries, combined with improving water quality, will lead to a resurgence of oysters. Both Maryland and Virginia saw oyster harvests reach thirty-five-year highs in recent years, yet the fishery is still a shadow of its former self. A full generation after crashing, oyster populations seem to be recovering, but could the crash have been avoided? No one knows what would have happened if Maryland or Virginia had implemented an oyster ban in the early 1990s. It is quite possible that MSX and dermo would have killed off oysters anyway, or that sediments and eutrophication would have smothered beds that otherwise would have been harvested. Still, haunting questions remain about what might have happened if a ban had been put in place to give existing oysters a few years' respite to reproduce. Would oyster populations in Maryland and Virginia have been further along in developing resistance to disease? Would a larger adult population have been able to take advantage of good environmental conditions to reproduce quickly, thus augmenting later oyster restoration efforts? Would oysters be celebrated alongside striped bass as a major success story? Or would a side effect of such strong medicine have been to kill off the fishing communities everyone involved agreed must be preserved? Of course, these questions are unanswerable, but asking them serves as a reminder of the long-term stakes when making difficult decisions about the Bay's health.

PREVENTING POLLUTION: THE PHOSPHATE BANS

While one goal of Chesapeake environmentalism was to protect the Bay's living resources, another was to reduce and eliminate pollution, which was ruining the Bay's water quality. Although many forms of pollution impaired the

Bay, nutrient pollution was "the Bay's public enemy number one." Nutrient pollution came from myriad sources, but once in the Bay, excessive amounts of nitrogen and phosphorus fueled algal blooms that blocked sunlight from reaching underwater grass beds. Worse, as these microscopic plant plankton died, their decomposition contributed to a biological process called *eutrophication*, which reduced oxygen levels in the Bay's waters, making it unfit for most forms of life. Chesapeake environmentalists combatted nutrient pollution on multiple fronts, but one of the most important and well-documented campaigns was the state-level effort to ban phosphates in laundry detergents throughout the watershed. The availability of alternatives to phosphates in detergents, the comparative ease of addressing the point sources of phosphorus, and the fact that the ban would reduce the amount of taxpayer money spent on operating sewage treatment plants made a phosphate ban seem like low-hanging fruit for Chesapeake environmentalism. However, the soap and detergent industry put up a vigorous and well-funded resistance to the phosphate ban. Citizen activism, led by CBF's volunteer corps, dubbed "the BayWatchers," helped prevail over industry lobbyists, and the resulting bans prevented millions of pounds of nutrient pollution from reaching the Chesapeake. Furthermore, the crucial victory in Maryland proved to be a tipping point for phosphate bans nationally. After other watershed states followed suit and environmentalists across the country agitated in their own states for a phosphate ban, the industry "ultimately decided it was cost effective to phase out the use of phosphorus in domestic laundry detergent." Once again, the Chesapeake's regional environmentalism would have national significance.[28]

The campaign to ban phosphates in Maryland during the 1985 state legislative session showed how quickly the political power of Chesapeake environmentalists had grown. Just a few years earlier, in 1981, a bill introduced to Maryland's General Assembly to ban phosphates had not made it out of the House Environmental Matters Committee following an 18–1 vote against the measure. Revisiting the issue would be a test of the movement's strength because the ban was opposed by an array of industry groups, ranging from national trade groups like the Soap and Detergent Association to local associations like the Maryland Chamber of Commerce. These forces financed a "small army of lobbyists who have turned the phosphate bill into one of the most heavily lobbied—and expensively fought—measures to pass through the General Assembly in many years."[29] Public disclosures indicated that opponents spent at least $650,000 to fight the phosphate ban, in addition to the baseline salaries already paid to lobbyists working for industry associations. Indeed, 1985 was the most expensive lobbying year in Maryland's history, and money spent to oppose the phosphate ban accounted for more than 10 percent of the total amount of money spent on all lobbying during the General Assembly session. By contrast, CBF spent $369,000 in 1985 for its entire environmental defense program, of which the campaign to ban phosphates was just one of many activities. CBF's William Baker said at the time, "I've never faced this type of money behind the lobbying effort of any bill I've worked on."[30] The chair of Maryland's House Environmental Matters Committee, Larry Young, agreed with Baker. He told the *Washington Post* that "the pressure has been awesome" and "the lobbying has been the most difficult I've undergone in my 11 years here."[31]

Opponents' strategy to combat the phosphate ban centered on playing up economic fears while trying to deny environmental benefits. Lobbyists from soap manufacturers claimed that a phosphate ban would cost consumers in Maryland $15 million a year and suggested that it would be ineffective. One lobbyist argued that "there isn't a shred of evidence that a phosphate ban would improve water quality one bit in the Chesapeake Bay."[32] The most outlandish claim came from a lobbyist who said that Baltimore City would lose more than a thousand jobs as a result of the ban. Investigative journalism by the *Sun*'s Tom Horton, who interviewed representatives of companies that might be impacted and learned that at most fifteen to twenty jobs might be affected, forced a retraction. Press coverage as a whole was generally supportive of the phosphate ban; the *Baltimore Sun*, the *Washington Post*, the *Capital* (Annapolis) and a host of smaller local papers all wrote editorials in favor of the law.[33]

Positive media coverage was crucial to the eventual success of the measure, but it was not a sufficient counterweight to the all-out effort of industry lobbyists, which was so great that at least one state legislator complained that he couldn't even escape them when he went to the bathroom.[34] It was concerted, grassroots citizen activism, led by CBF's BayWatchers, that got the bill passed. CBF staffer Ann Swanson organized the BayWatchers. It helped launch her career, which eventually took her from CBF to becoming the longest-tenured director of the Chesapeake Bay Commission. In 1985 the BayWatchers were a network of eighteen hundred CBF members who had signed up to "be mobilized on short notice to become involved in activities that could have a significant impact on the cleanup of Chesapeake Bay."[35] Swanson made the BayWatchers into a force by arming motivated citizen activists with the information and skills to persuade their elected representatives. Swanson sent out "Action Alerts" containing accurate, easy-to-understand scientific information and precise instructions on how to call or write elected representatives, write letters to the editors of newspapers, and speak at a public hearing. A handout titled "Your Right to Write" is typical. The handout instructed members to "be courteous, patient, appreciative, and realistic," while still "being firm and politely persistent." The handout also provided useful tips, such as to include the bill number, write to the appropriate legislator, and above all keep up the pressure. The handout told members to "ask your representative to state his position when he replies. If your elected official is equivocal in his response, write again and request clarification. Don't hesitate to ask questions. However, don't sound demanding or threatening."[36] Swanson complemented the "Action Alerts" with frequent workshops and meetings to keep the BayWatchers organized, motivated, and effective.

Swanson's organizational skills and CBF's reputation were direct targets of opponents of the phosphate ban, who tried to undermine the organization's efforts by forming an "astroturf" group to sow confusion. In February 1985 an organization calling itself the Consumers League for Environmental Action Now (CLEAN) registered in Maryland as an official lobbying group. CLEAN, a classic example of an "astroturf" group, billed itself as an environmental group, yet it received all its funding from the Coin Operated Laundry Association, and its president owned a coin-operated laundry. CLEAN distributed pamphlets and brochures that mimicked CBF's blue-and-white "Save the Bay" logo so that unsuspecting citizens might think that CLEAN was affiliated with CBF. CLEAN's misdirection campaign included a petition that claimed to be pro-Bay. Hidden beneath several generic proenvironmental claims about saving the Bay and reducing nutrients was a statement opposing the phosphate ban because it was not "the right way" to clean up the Bay. In addition to mimicking CBF and disguising its true goals, CLEAN offered volunteers one dollar for each signature they collected. By March 1985 nearly two thousand people had signed CLEAN's petition.[37]

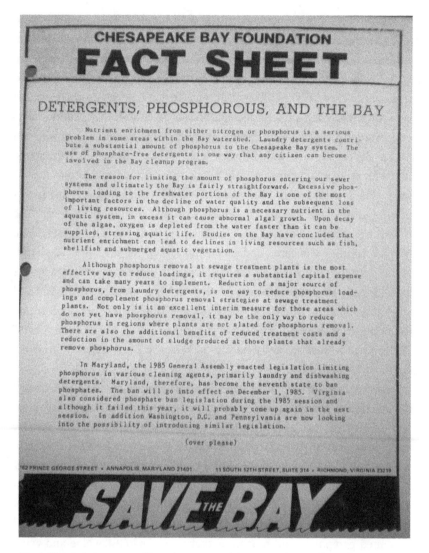

"Fact Sheet: Detergents, Phosphorous, and the Bay," 1985. (Photo by author, personal collection of Ann Swanson)

CLEAN's dirty tactics came to light in March 1985 when Molly Broderson realized that she had made a terrible mistake. As she recalled, "After collecting nearly eighty signatures, two individuals refused to sign the petition and informed me that I was working contrary to the efforts of the environmental community. . . . After talking to these individuals, I immediately called the Chesapeake Bay Foundation." Broderson talked directly to William Baker and learned "with much embarrassment" that she had been duped by CLEAN. With Baker's and Swanson's help, Broderson struck back at CLEAN, first by publishing a letter she wrote to everyone who had signed her petition informing them of what CLEAN actually stood for and apologizing for her error. "I thought I was giving you an opportunity to participate in the cleanup," she wrote. "I have since come to find out that they [CLEAN] are part of the opposition." Her letter went out to every CBF BayWatcher, and her story helped combat the efforts to undermine grassroots organizing on the issue. Broderson appeared at press conferences, talked to television and print reporters, and later testified before the Maryland House of Delegates.[38]

Broderson's story came to light at a pivotal moment in the legislative session. A bill to ban phosphates had passed the Maryland State Senate by vote of 39–5, but it faced a much stiffer challenge in the House Environmental Matters Committee. Gerald Winegrad, a leader on environmental issues in the Maryland Senate, remembered being "taunted by antagonistic members of the House of Delegates when my ban on detergent phosphates passed the Senate. They jeered, 'Now, let's see you get it out of the House' and tried their best to block the legislation." Perhaps if Molly Broderson hadn't been corrected by two friends of the Chesapeake whose identities have been lost to history, and if she hadn't had the introspection to check with CBF and the courage to admit she was wrong, the CLEAN front would have succeeded in sowing enough confusion to allow members on the Environmental Matters Committee to vote no and claim that environmentalists themselves were divided on the issue. Instead, CLEAN's tactics backfired, and a raft of negative press coverage combined with steady pressure from the public, including Broderson's star-witness testimony before the committee, forced action on the bill, thwarting the opponents' plan to stall until the legislative session ended and claim that there had not been clear enough information or consensus for a vote. Rather than accepting the Senate's bill, the House Environmental Matters Committee sent its own, heavily amended version of the bill to the House floor, hoping it would be voted down. One lobbyist bragged that "a lot of vodka went into those amendments," and the *Post* called it "a tactical victory" for the industry's lobbyists.[39]

The industry's so-called victory ended up being a strategic error. Swanson sprang into action, calling upon the BayWatchers to tell their legislators to vote yes and fix the bill's issues in conference committee. Swanson told the BayWatchers that "the bill was weakened in the house committee in an effort by opponents to make the bill hopelessly unworkable. They hope that the House now will *not* vote for it in its unworkable state. We need to counter this threat with phone calls telling Delegates to vote YES so that the bill is *passed*. If passed, the bill will go to conference committee and the weakening amendments will be eliminated or toned down." Swanson provided a step-by-step guide on how to contact one's own legislator, as well as instructions and a guide for persuading friends and acquaintances to do the same. Her plan included a coordinated push for volunteers to call their delegates the morning of the vote and tell them "that they have heard of *no* problems associated with non-phosphate detergent" and that "they recognize that Phosphorus is 1 of many Bay problems, but this is a way they can make a contribution to the Bay's health." With well-coordinated grassroots pressure, the bill passed the House of Delegates, and although it took until the final day of the state's legislative session, the phosphate ban became law. The only remaining amendment was a sunset provision that would end the ban after four years unless the General Assembly took further legislative action, which it did in 1988, repealing the sunset provision.[40]

Maryland's phosphate ban had an immediate, beneficial impact on the Bay. An initial review of the ban's effects found that "within only a few months of banning phosphates in detergents, the phosphorus concentration in wastewaters generated within the WSSC service district [which served large chunks of the DC suburbs] dropped by 32%." A more thorough study by the state in 1987 concluded that the ban had saved taxpayers more than $4 million by easing the chemical burden on wastewater treatment plants. As the soap industry had feared, after Maryland's ban the dam burst. Washington, DC, banned phosphates in 1986, and after "an intense lobbying campaign by both the soap and detergent industry and numerous environmental and consumer groups inside and outside the state," Virginia banned phosphates in 1987. According to CBF's Virginia director, Joseph Maroon, "The positive results from the Maryland and D.C. bans" and "the broad grassroots support throughout the state," combined with support from Virginia's governor, Gerald Baliles, made the phosphate ban possible. Pennsylvania was a tougher nut to crack, but it too banned phosphates in 1989. It was a rare win-win that was both more economically cost-effective and environmentally friendly.[41]

The phosphate ban represented one of the successes of Chesapeake environmentalism, but it also revealed its limits. Considering that the ban was good economics and good environmentally, it is stunning how long it took for the states to pass their bans. The pessimistic reading of this history is that for environmentalists to pluck even the lowest-hanging fruit—a straightforward law that immediately and directly improved the Bay's water quality *and* benefitted the bottom line of state governments, all while requiring minimal industry sacrifice—they have to be prepared for a bruising fight. As the CLEAN example shows, even on a relatively trivial issue industry groups were willing to adopt what Naomi Oreskes and Erik Conway have called "the Tobacco Strategy" of "fighting facts, and merchandising doubt." Chesapeake environmentalism would face even more formidable opposition to further-reaching laws that required greater regulation and personal sacrifice. However, the pessimistic reading of the phosphate bans is not the whole story. When Ann Swanson's BayWatchers helped get the phosphate ban passed in Maryland, it immediately upped the pressure on lawmakers in DC, Virginia, and Pennsylvania, cosignatories with Maryland to the Chesapeake Bay Agreement. The regional alliance to save the Chesapeake created a dynamic according to which a breakthrough in one state increased the odds of breakthroughs in the other states. This dynamic was revealed in the striped bass and phosphate bans, and even more powerfully by its absence in the effort to protect the oysters. The regional nature of Chesapeake environmentalism meant that while there were many fronts Bay protectors needed to defend, there were also many opportunities to go on the offensive.[42]

EXPANDING CHESAPEAKE ENVIRONMENTALISM TO PENNSYLVANIA

The single biggest chunk of the Chesapeake Bay's watershed, roughly 35 percent, lies within the Commonwealth of Pennsylvania. Given that none of the estuary itself lies within the commonwealth, Pennsylvania presented a unique challenge for Chesapeake environmentalists. Getting Pennsylvania to join the Chesapeake Bay Agreement in the first place was a major coup, but getting the commonwealth to take meaningful steps toward reducing pollution would require a major investment of time and resources. In 1985, after reflecting on the first wave of accomplishments following the Chesapeake Bay Agreement and planning next steps, CBF's William Baker wrote that "first and foremost, we will seek greater involvement from the Commonwealth of Pennsylvania." The foundation would tackle many issues in Pennsylvania, but the most significant target in its crosshairs was the millions of pounds of nitrogen, phosphorus, and sediment flowing from fertile Pennsylvania farms down the Susquehanna River and into the Bay.[43]

CBF's first step toward greater involvement in Pennsylvania was to establish a permanent office in the state capital, Harrisburg. Thanks to a grant from the Pittsburgh-based R. K. Mellon Foundation, CBF opened its first satellite office in Harrisburg on October 30, 1986. Ann Powers, CBF's vice president and general counsel, hailed the office as a major milestone "because it represents a new phase with our efforts on the Bay. The Chesapeake Bay Foundation has become a truly regional organization." Powers found CBF's reception in Pennsylvania "gratifying," and she believed there was "a large reservoir of interest in the Bay and good will toward those who are working on its problems." Tapping into that reservoir of interest would prove to be more difficult than anticipated. Powers recognized that one of the biggest challenges facing the group was that "when most people think of the Bay, they don't think of manure management in the Pennsylvania Dutch country." Linking the health of the Bay to the health of Pennsylvania farms would be challenging, especially when the former required sacrifices of the latter. Although CBF's top priority in the state was to reduce agricultural nutrient pollution, it would take the organization seven years, with several false starts, to build a coalition strong enough to pass legislation, the Nutrient Management Act, which would limit manure in the Pennsylvania Dutch country, among other places and pollutants. Progress was slow in part because CBF had not built up as much grassroots support in Pennsylvania as it had in Maryland and Virginia. In June 1985, a year before opening its Harrisburg office, the foundation had only four thousand members from Pennsylvania. However, the bigger challenge was one of political ecology: Pennsylvania was a major upstream source of pollution, but without any of the Bay waters lying within state lines, it would not directly benefit from a cleaner Bay. In its press release, CBF called opening a Pennsylvania office "a long delayed dream";

the question in the late 1980s was whether an enduring presence for CBF in Pennsylvania would become a reality.⁴⁴

The man tasked with leading the charge in Pennsylvania was Thomas P. Sexton III, who did much to build CBF's capacities in Pennsylvania but would leave the organization before his efforts came to fruition. Sexton was a Pennsylvania native whose varied career included environmental education in the Poconos and Hawk Mountain, natural resource management on the Schuylkill River for the National Park Service, and just before taking the CBF post, a position as a senior congressional staffer. Sexton's challenge was to figure out how to make Pennsylvanians care about the Bay and see their connection to it. He argued that "how we, in Pennsylvania, treat the Bay comes right back to us in recreational opportunities and in the food products that we enjoy. A healthy Bay is in everyone's best interest." Sexton's reasoning shows the limits of CBF's approach. How would a healthy Bay benefit a Lancaster County dairy farmer who did not have the time or money to go sailing on the Chesapeake or purchase a bushel of crabs? As in Maryland and Virginia, environmental education would lead the way in Pennsylvania by connecting the Bay to people's backyards. On May 18, 1987, Sexton took the first crew of Pennsylvania students on a "Chesapeake" field trip in their home state. Students from Camp Hill High School and a representative from Pennsylvania's Department of Environmental Resources took an eight-mile canoe trip on Conodoguinet Creek, a tributary that winds sinuously for more than a hundred miles through central Pennsylvania farm country before emptying into the Susquehanna at Harrisburg. The trip earned rave reviews; Bill Kimmich, a biology teacher, gushed to Harrisburg's *Patriot-News* that "in one afternoon on the creek, they learn more than nine months in a classroom, and it's more meaningful." With its education program allowing the organization to make inroads in the state, CBF would triple its Pennsylvania membership by 1989 and establish something of a beachhead for Chesapeake environmentalism in the south-central part of the state.⁴⁵

While Sexton was earning some goodwill and good press with the education program, he had less success with his first foray into agricultural policy. CBF as a whole made a strategic move to focus more resources on agricultural pollution across the watershed at the end of 1986 by hiring Patrick H. Gardner, an attorney with a background in agricultural economics who had previously clerked for the U.S. Department of Agriculture and the Sierra Club Legal Defense Fund. This was not CBF's first experience with agriculture; the organization had owned and operated a demonstration farm, Clagett Farm, as part of its educational program in Maryland since 1981; however, Gardner's hire represented a realization that agricultural nutrient pollution needed to be prioritized across the watershed. The agricultural policy initiative Gardner was to lead had three major prongs: improving the management efficiency of existing state and federal agricultural pollution control policies; providing incentives for individual farmers to adopt and implement best practices; and establishing new laws and policies to regulate agricultural pollution. According to Gardner, "The agricultural project embraces the philosophy that voluntary programs encouraging good civics and smart economics are preferable to mandatory controls. Mandatory controls are a last resort, limited to bad actors, or as an alternative strategy should voluntarism prove ineffective for solving the Bay's agricultural pollution problems." This philosophy would prove naïve and idealistic. While Gardner's vision ran into problems across the watershed, Pennsylvania would give him, Sexton, and CBF a hard lesson in agricultural realpolitik.[46]

In 1987 Sexton focused on the first two prongs of CBF's agricultural policy, government efficiency and supporting individual farmers. To that end, his main priority was supporting a statewide referendum that would make $100 million in bonds available for farmland preservation and provide incentives to encourage the use of pollution control practices on farmlands preserved under the program. One virtue of this program was that it would help protect farmland from development. As concerned as CBF was about agricultural pollution, it saw that as a controllable problem and greatly preferable to the expansion of suburbs and strip malls. The other virtue was that the proposal already seemed likely to garner widespread support, so advocating for the referendum would be an easy win that would allow CBF to claim a feather in its cap and demonstrate its support for farmers. Indeed, protecting Pennsylvania's agricultural heritage was a popular cause, and the referendum was approved by voters in a two-to-one landslide. This policy would help prevent future pollution from urban sprawl, but it did little to immediately reduce nutrient, sediment, or chemical pollution from Pennsylvania farms. For that, Sexton and Gardner turned the following year to the federally funded Conservation Reserve Program (CRP), which Congress had established as part of the 1985 Farm Bill. CRP was a voluntary program that paid farmers to retire erosion-prone portions of their land for ten years. There was great hope as late as 1988 that CRP "may well become the single most important agricultural water quality program in the 'Save the Bay' campaign," but only 5 percent of eligible lands in the watershed were enrolled in the program. Despite their best efforts, Sexton, Gardner, and the rest of the CBF team reported a year later that CRP and similar programs were not effective tools because "the majority of farmers do not participate in these programs, and those who do participate are usually already using BMPs [Best Management Practices]." After two years, the first two prongs of CBF's agricultural policy had proven mostly ineffective. That left them with their option of last resort, mandatory controls.[47]

William Baker signaled in early 1989 that CBF would have to take a more forceful approach. Expressing frustration with the pace of the Bay's recovery and the ongoing problems it faced, Baker proposed a major philosophical shift in the organization: "We will abandon the concept that somehow individual companies, farmers, municipalities, or homeowners will voluntarily go the extra mile to protect the environment." Brushing off the last vestiges of Arthur Sherwood's philosophy of environmental reasonableness, he called for something that would have appealed to Jess Malcolm, "a structure of environmental laws and regulations that are both strict and equal for all sectors of society." On the Pennsylvania agricultural front this meant lobbying for the Nutrient Management Act, the key features of which would require farmers to develop nutrient management plans to reduce manure and fertilizer runoff and require the state to take action against violators. This was a departure from previous efforts to reduce agricultural nutrient pollution, and CBF recognized that "to ensure passage and implementation of legislation of this complexity will take years." It would also require assembling a broad coalition of support and getting some degree of buy-in from farmers and farming communities.[48]

To that end, CBF sought allies and wooed members of the Pennsylvania agricultural community. One of the foundation's most important allies was David Brubaker, of Lancaster, who in addition to having a prominent industry role as the executive vice president of PennAg Industries Association and PennAg Industries Service Corporation was also a member of the Citizens Advisory Committee of the Chesapeake Bay Program. Brubaker believed that "agriculture, or any type of business, is not sustainable over time if it is not socially responsible and if it is not environmentally sound." For being "a courageous champion" of responsibility toward the environment, CBF recognized him as their 1989 Conservationist of the Year. In addition to Brubaker, CBF secured bipartisan sponsorship from Democrat Jeffrey Coy and Republican Noah Wenger. Both men represented rural, agricultural districts, and their support was key because a farm bill driven by legislators from Pittsburgh or Philadelphia would have been dead on arrival. Ann Swanson, who by this time had left CBF to become the executive director of the Chesapeake Bay Commission, remembered that in addition to their genuine concern for the environment, both men saw a tactical advantage for Pennsylvania in choosing to clean up agriculture on its own terms rather than doing nothing and risking a federally mandated cleanup. In addition to CBF's growing staff of agricultural policy experts (which had expanded to include Lamont Garber, a Penn State grad with a bachelor of science in agricultural economics), the organization teamed up with the Rodale Institute and the Penn State Agricultural Extension for their technical expertise in agricultural matters. Swanson also credits George Wolff, a farmer and lobbyist, with building grassroots support through an informal network called "The Group That Didn't Exist," which met for breakfast strategy sessions. The need for such a furtive approach to lobbying for the bill illustrates how much of an uphill battle it faced. Nevertheless, with the help of a broad coalition of local allies and supporters, the Nutrient Management Act became more than just a demand from an out-of-state environmental nonprofit; it was a thoughtful policy proposal that grew organically from the state's own concerns and commitments. Chesapeake environmentalism had arrived in Pennsylvania at last.[49]

Before achieving a meaningful victory, Chesapeake environmentalism would have to endure a humiliating setback in Pennsylvania, even though a breakthrough seemed imminent. In 1990, Governor Robert Casey Sr., a Democrat, created a Select Committee on Nonpoint Source Nutrient Pollution, which included Jeffrey Coy, George Wolff, and William Baker. Unsurprisingly, the committee recommended that farmers develop and implement nutrient management plans with state funding for implementation and state enforcement for bad actors. Governor Casey signaled his support, claiming that he was "committed to embarking on a new effort to address its recommendations." Unfortunately for the bill's supporters, not many in the state senate shared Casey's commitment. Despite passing the Pennsylvania house in mid-1991, house bill 496, A Nutrient Management Act, stalled completely when it reached the senate. There the bill languished for the remainder of the 1991–92 legislative cycle. It was an embarrassing defeat, one that caused some to cast doubt on the entire project of Chesapeake environmentalism. What happened?[50]

The most acerbic account of what went wrong was by the longtime chronicler of Bay environmental news Tom Horton of the *Baltimore Sun*. Calling the bill's failure a "fiasco" and a "bitter defeat," Horton highlighted a literal barnstorming tour as the bill's undoing. According to his reporting, Allen Weicksel, a dairy farmer and retired Army Special Forces officer, started a grassroots group called the Family Farm Movement. He "criss-cross[ed] the state, holding 102 meetings between March and May in barns and kitchens, rousing opposition to the bill among the Amish and other farmers." Weicksel believed that his previous career with the Army as a "special adviser" had prepared him for just such a role; he said that "organizing this movement was just an extension of what I've done my whole career." What Weicksel did was bring the progress of a fairly widely supported bill to a grinding halt. At a key juncture, "Weicksel managed to dominate a Senate agricultural committee hearing on the bill, showing up with more than a hundred farmers, including dozens of Amish, who very seldom participate publicly in such affairs." Four years of work building a coalition between environmentalists and the agricultural community unraveled over the course of several weeks as the senate committee first weakened the bill so that it applied to only 15 percent of Pennsylvania farms and then ultimately refused to vote on the bill before the legislative session ended. It led Horton to send a not so thinly veiled shot across CBF's bow, asking whether the debacle "ought to cause some rethinking among a number of save-the-bay organizations that have relied heavily on consensus-building."[51]

Horton's reporting highlighted the theatrics of Weicksel, but there were many less dramatic voices who questioned the impact of such a law on the economics of farming. For instance, CBF's Garber acknowledged that "the ability to make a profit is just as important as environmental protection." Radical leftists they were not. To the organization's credit, it refused to see farmers as enemies of environmental protection, and staff redoubled their efforts to build support for the legislation, which would take another two years.

CBF was not able to get traction on nutrient management legislation in 1992, but the group sponsored a sustainable agriculture conference, which they successfully spun off into a new partnership, the Pennsylvania Association for Sustainable Agriculture. The following spring, CBF invited members of farmers' associations to a farmers' summit at their Port Isobel Island education center to try to find common ground. Robert Hoyt, who had taken over for Sexton as the top CBF staffer in Pennsylvania, said, "The leaders of the farm organizations deserve a lot of credit for coming to the table. . . . They're in a tough position because they want to protect the environment, but they can't risk going out of business because of regulations. We agree. We don't want to see farmers go under. . . . Family farms are a part of the beauty and culture of this region. We don't want to lose that. So when you realize that both groups want a clean environment and both groups want to maintain farms, you can craft a program that serves those two goals." Weicksel's antics had managed to create a costly and embarrassing delay, but he could not stop the momentum building toward action on nutrient management. The reinvigorated coalition of farmers and environmentalists, combined with Governor Casey's support for the bill after winning reelection in 1992, persuaded legislators to reach a deal for swift passage of the Nutrient Management Act in 1993. Although the bill was long delayed, legal scholars noted that once passed, it made Pennsylvania "the first state in the Chesapeake Bay watershed and one of the first states in the nation to adopt mandatory nutrient management controls on farm pollution."[52]

In the seven years after CBF opened its office in Harrisburg, the group had managed to build a coalition to pass a major law and in so doing had secured a public declaration of support from Governor Casey. "Pennsylvania will continue as full partners with Maryland, Virginia and the District of Columbia," he promised, "so that the Chesapeake Bay remains an important economic, environmental recreational and aesthetic resource for future generations." However, the struggle to pass a strong agricultural pollution law in Pennsylvania illustrates another recurring dynamic in Chesapeake environmentalism. Simply put, Pennsylvania is a junior partner in the movement. For structural reasons ranging from the absence of Chesapeake shoreline within state limits to the more conservative political lean of the state, Pennsylvania has consistently lagged behind Maryland and Virginia in Chesapeake issues. In a 2022 interview, Hilary Harp Falk, who succeeded William Baker as president and

CEO of CBF, said, "With respect to Pennsylvania, we've been concerned for a while that while Maryland and Virginia in particular have made significant progress and investments in clean water that Pennsylvania was behind. Significantly behind." In 2004 Baker complained that "the train is leaving the station" and "Pennsylvania is running to jump on." In 1988 CBF ran an editorial demanding, "Pennsylvania needs to bolster its commitment. Although Pennsylvania does not border the Chesapeake, it contributes significantly to the pollution of the Bay and should be a fully committed partner to the cleanup." Before the 1983 Chesapeake Bay Agreement, CBF was so fearful that there would be no commitment from Pennsylvania that it called for Maryland and Virginia "to seriously consider funneling some of their own states' funds for Bay clean-up to Pennsylvania." In addition, as we've already seen, Pennsylvania was the last of the three states to implement a phosphate ban. For more than forty years, Pennsylvania has lagged behind Maryland and Virginia.[53]

Yet, the conclusion here is not a bleak one. Rather, the difficulties Chesapeake environmentalists faced in Pennsylvania must be understood within the state's historical context. As the history of CBF's expansion into Pennsylvania shows, the difficulty of getting the state to commit to major actions should not be underestimated. The state lacks the same economic incentives and cultural heritage as Maryland and Virginia. In that context, Pennsylvania, even as a junior partner, matters a great deal *because it is still a partner*. Where Maryland and Virginia lead, Pennsylvania will eventually follow, especially if there is prodding, or the threat of prodding, from the U.S. EPA. In some cases, as with agricultural pollution, Pennsylvania can even be a bit of a leader. This is far from an ideal situation for Chesapeake activists, but Pennsylvania's continued involvement as a critical partner to the Bay Program has been key to the movement's political and ecological sustainability.

CONCLUSION

Rather than one single narrative, this chapter has looked at four case studies of Chesapeake environmentalism in the decade after the first Chesapeake Bay Agreement. Many important issues, such as legal challenges that went to the U.S. Supreme Court, land preservation, toxic pollution, environmental education, and more, have had to be left out in the interest of space. However, one final episode from this decade brings together many of the themes from this chapter and shows critical changes in the movement. In 1987 and again in 1992 the states and the federal government (along with the District of Columbia and the Chesapeake Bay Commission) renewed their commitment "to restore and protect the ecological integrity, productivity and beneficial uses of the Chesapeake Bay." These successive pledges, which have been criticized as so much hot air, deserve a second look.[54]

By 1986, all the officials who had signed the initial Chesapeake Bay Agreement either were out of office or soon would reach their term limits, except for DC mayor Marion Barry. Likewise, the looming retirement of the Bay's greatest congressional champion, Senator Charles Mathias, raised questions about who would lead the charge at the federal level. Worst of all, on June 1 the *Washington Post* published a bombshell piece of investigative reporting, "The Poisoning of Chesapeake Bay," which all but called the restoration movement a failure. After the initial adrenaline rush following the first Chesapeake Bay Agreement, CBF reported that in some circles "a complacency has settled in, resulting in a belief that everything that can be done is being done." Recognizing his iconic role in the movement, Senator Mathias convened a Senate subcommittee hearing on the future of the Chesapeake Bay Program on June, 24 1986. His goals were to help refocus attention on the Chesapeake Bay cleanup and again, as in 1983, to to push the EPA administrator, this time Lee Thomas, to commit to a new agreement that would have more specific goals. Notably, Mathias received some pushback from William A. Cook, who was at the hearing representing Pennsylvania. Cook said, "Pennsylvania believes that we have a good coordinating mechanism in place with the executive agreement signed in 1983 and believes it should be given full opportunity to continue the good work that it has done at this stage." Despite some objections from the northernmost partner in the Bay cleanup, in the summer of 1987 the southernmost partner, Virginia, under Governor Gerald Baliles, scheduled another Bay Summit, which led to the second Chesapeake Bay Agreement.[55]

Signed on December 15, almost four years to the day after the original, the 1987 Chesapeake Bay Agreement committed the signatories to a 40 percent reduction in nutrient pollution by 2000. Although the agreement still did not include the other three watershed states—New York, Delaware, and

West Virginia—Pennsylvania remained in the fold, and the commitment to a specific target led the *Post*'s editorial board to praise the agreement, calling it "a pledge of life for the Bay." In a narrow sense, this agreement was a failure. The year 2000 came and went without the states hitting their 40 percent nutrient reduction targets. But in a year that began without phosphate bans in Virginia or Pennsylvania, in a year when Maryland watermen still couldn't catch striped bass, the renewed commitment to the Bay allowed environmentalists to press new political leaders for policies to protect the Chesapeake. For example, if Pennsylvania's Governor Casey had not signed the 1987 agreement, CBF would have had less leverage during the arduous process of passing the Nutrient Management Act. The 1987 agreement set a precedent that new generations of political leaders would personally commit to restoring the Bay. This dynamic repeated itself five years later, in 1992, with a set of amendments to the 1987 agreement designed to focus on the Bay's tributaries as well as the estuary itself.[56]

Clearly, by the 1990s the Chesapeake environmental movement had blossomed across the watershed and established itself as a permanent part of the political landscape. Yet, the man who was arguably the region's most prominent environmentalist was not happy. After the 1992 amendments, William Baker wrote to the signatories that "dramatic actions must be taken" or else "progress made to date may be lost." Baker and others had hoped for more comprehensive plans, specific goals, and most of all, accountability. Yet Baker was "disappointed" in the final 1992 amendments because "where commitments for action have been made, it is frequently impossible to determine who is accountable." For Baker and likely for many others, it was "enormously frustrating." By the 1990s, Chesapeake environmentalists had made real progress and cemented their place as a powerful political force in the region. As they did, however, they consistently encountered a backlash, often from parties opposed to environmental policies because they threatened them economically. Commercial watermen, farmers, and even the laundry detergent industry had practical reasons to oppose environmental measures for the real or perceived harm they could cause. Soon, those with such practical concerns would be joined by a zealous, ideological opposition to environmentalism driven in large part by the nationwide backlash among Republican politicians opposed to nearly any form of government regulation. This backlash coincided with fatigue and disappointment among Chesapeake environmentalists that their movement had not yet achieved its goals after a decade of labor. Baker's disappointment and frustration with the 1992 amendments reflected the reality of this emerging opposition to saving the Bay. Overcoming that opposition would prove to be even more daunting and time-consuming than getting the regional partners to commit to a joint approach to the Chesapeake's recovery had been in the first place.[57]

FIVE

MAKING A NATIONAL TREASURE

The Chesapeake Bay's environmental advocates have argued throughout the movement's history that the Bay's ecological, cultural, and historical significance is not just regional; it is national. In 1968 Jess Malcolm, CBF's star-crossed first executive director, said that protecting the Bay was "truly a regional, if not a National issue." In 1987, the second Chesapeake Bay Agreement's very first sentence proclaimed, "The Chesapeake Bay is a national treasure and a resource of worldwide significance." State and federal officials have echoed these lines at least as far back as the 1933 conference on the Bay, with nearly every major political figure in this narrative agreeing with the basic premise. Despite the rhetoric, these same actors continued to treat the Bay as more of a regional concern. By the mid-1990s, the failure of the federal government to match the rhetoric of its officials with legislative and financial support was especially glaring. In 2003 the Chesapeake Bay Commission found in an independent review that federal funding covered only 18 percent of the cost of Bay protection and cleanup in the two decades following the first Chesapeake Bay Agreement. By contrast, federal funding covered roughly 31 percent of Chesapeake spending in the period 2015–23. While the state of Maryland remains the leader in Chesapeake spending, this chapter chronicles the federal government's growing and essential role in funding the Bay's recovery.[1]

After fitful but meaningful progress in the decade following the first Chesapeake Bay Agreement, the movement faced setbacks in Annapolis, Richmond, and Harrisburg, but most especially inside the Capital Beltway, where ideological opposition to environmental protection by conservative Republicans threatened to undermine the Bay's fragile recovery. By the mid-1990s, the momentum to protect the Chesapeake had utterly stalled. What rescued the movement from its doldrums was an environmental disaster caused by a mysterious microorganism, *Pfiesteria piscicida*. Just as Tropical Storm Agnes had catalyzed the regional environmental movement in the 1970s, the *Pfiesteria* disaster reignited the movement in the 1990s, leading to another Bay agreement, dubbed Chesapeake 2000. *Pfiesteria* made Chesapeake protection a top political issue again, and at the state level Republicans and Democrats worked together to push funding levels for Bay protection to their highest levels ever. A similar federal response was notably absent under the George W. Bush administration, which made the antienvironmentalism of conservative Republicans into an art form. Nevertheless, early in the twenty-first century the movement to protect the Bay had reached a pivotal juncture. When the political winds shifted with the election of Barack Obama, Chesapeake environmentalism seized the opportunity to force the federal government to fulfill its commitments to the Bay, or in Obama's words, "to begin a new era of shared Federal leadership with respect to the protection and restoration of the Chesapeake Bay."[2]

THE MID-1990S: CHESAPEAKE ENVIRONMENTALISM AT LOW TIDE

Chesapeake environmentalists faced setbacks throughout the region, but the symbolic low-water mark came on October 9, 1995, when orange flames and black smoke shattered the nighttime tranquility of Tylerton, a small fishing village nestled in the heart of the Bay. Located on Smith Island, a small archipelago of marshes straddling the Maryland-Virginia border, Tylerton was home to a CBF educational center that had gone from being a point of pride in the community to a literal spark of division. No one was ever held responsible for the arson that caused thirteen thousand dollars' worth of damages and, more significantly, destroyed the sense of trust and shared purpose that had once characterized relations between environmentalists and islanders. "Tylerton embraced us," recalled Don Baugh, CBF's former vice president for education, who helped set up the education center in the village. Baugh's partner in Tylerton, William Goldsborough, who started as a CBF educator, wrote that "Smith Island people . . . have made us feel at home in Tylerton and on Smith Island and have been extremely receptive and friendly to our student groups. Several watermen have taken out individuals from our groups scraping and potting." Yet by 1995 many of those same watermen wanted CBF gone.[3] The Tangier Sound Watermen's Association hung a billboard across the channel from CBF's buildings that read, "SMITH ISLANDS [sic] WAY OF LIFE WILL SOON BE OVER DUE TO THE CHESAPEAKE BAY FOUNDATION PLEASE DO NOT SUPPORT THEM." There were more ominous warning signs before the fire—oil was dumped on a CBF education boat; a streetlight was shot out in front of the building that would later be burned; and a dead cormorant mysteriously appeared in a foundation aquarium. Although the fire on the night of October 9 was put out quickly, the harm it caused would take years to heal.[4]

The arson in Tylerton was just the most violent example of an antienvironmental reaction sweeping Chesapeake and national politics. Many observers have begun to analyze the causes and consequences of the rightward lurch in Republican politics that occurred by the 1990s, even as the ramifications from this period continue to unfold. As an addition to the emerging literature on this subject, the Chesapeake's history serves as a concrete example of how local and national environmental politics interacted at the end of the twentieth century. There is no dispute that after Reagan's failed bid to reverse the environmental gains of the 1970s, conservative intellectuals and politicians captured the Republican Party. As Judith Layzer explains, "Throughout the late 1980s and early 1990s, conservative think tanks disseminated an analytic framework that comprehensively dismissed pollution control regulation," an investment of resources and energy that by 1994 "paid off in Congress, where a newly installed Republican leadership launched a series of direct attacks on environmental regulation." Most famously, the conservative ideological framework challenged any action whatsoever on climate change, but the ideological whiplash from George H. W. Bush, who signed the 1990 Clean Air Act Amendments, to Newt Gingrich's Contract with America only four years later is worth remembering. Although it seems to have been baked into the DNA of the modern Republican Party, antienvironmentalism was a comparatively recent and fairly sudden development. As already discussed, Reagan's antienvironmental agenda was derailed in the Chesapeake by Maryland Republican Charles Mathias, among many others. However, by the 1994 midterm elections Mathias had long been retired, and bipartisan support for the Chesapeake at the federal level was as imperiled as the estuary itself.[5]

The 104th Congress quickly developed into a politically polarized legislative body, with the majority party antithetically opposed to environmental protection. Nolan McCarty found the 104th to be the most polarized Congress since World War II. After retaking Congress for the first time in a generation, GOP leaders were in no mood to cooperate. The newly elected Republicans included the most antienvironmental group of legislators in at least a generation. Debra Callahan, president of the League of Conservation Voters, called the 104th Congress "an environmental train wreck." For instance, the average LCV score for the incoming GOP class was 12 out of 100, with 36 members of the 73-member freshmen class, including 7 Republican legislators from Chesapeake Bay states, earning zeroes by voting against the environment every single time. Both represented a grim foreshadowing of the direction Republican politics would take.[6]

These new conservative legislators were more ideological and less willing to compromise than any lawmakers environmentalists had previously faced. Ideological fervor was not limited to the environment; the 1994 Republican cohort was "more extreme than their predecessors in their refusal to learn how to work in the political system or to build any kind of coalitions." Richard Andrews delivers what has become the consensus view: "This politics of the 'base,' of mobilizing and catering to the most zealous constituencies within each party . . . reflected a significant shift in American politics generally and particularly from the broadly bipartisan environmental politics of the 1970s. . . . In Congress, ever since Newt Gingrich and the 1994 elections the Republican leadership in both houses had pursued a purposeful and disciplined agenda of promoting business interests and weakening environmental protection policies, using all possible means including even systematic manipulation of science and language." Worse still for Chesapeake environmentalists, some of the most strident antienvironmentalists came from within the watershed. For instance, Bud Shuster (R-PA) was the main sponsor of the infamous Clean Water Amendments, dubbed the "Dirty Water Bill"; and Thomas Bliley Jr. (R-VA) pushed an antienvironmental deregulatory agenda from his position as chairman of the Commerce Committee. Such was the power of the political winds blowing the country at gale force to the right that CBF, like many environmental groups, tried to adjust its rhetoric to fit the times. For instance, in early 1995 the *BayWatcher Bulletin* reasoned that while "at first glance the vote for streamlining government may seem like a rejection of the Chesapeake Bay," in reality, "in most instances, what is good for the Bay is good for business and helps to reduce taxes." This was wishful thinking, and it was doomed to fail against opponents who took it as an article of faith that what was good for the environment was bad for the economy.[7]

Known officially as H.R. 961—Clean Water Amendments of 1995, the "Dirty Water Bill" highlighted the new reality of Republican hostility to environmental protection. It earned its moniker and made national headlines because of how thoroughly and brazenly it aimed to undo the Clean Water Act. Among a host of other problems with the bill that CBF called "an assault on water quality protection," in the Chesapeake watershed H.R. 961 would have stripped federal protection from 70 percent of the wetlands. The *New York Times* editorial board concurred, calling the amendments "crippling revisions to the Clean Water Act of 1972." The *Times* singled out Shuster for extra scorn, calling out as "appalling" his allowing industry lobbyists to write the bill that claimed to "regulate" them, while systematically excluding input from environmentalists or the EPA. When the House passed Shuster's measure in May 1995, President Bill Clinton immediately vowed to veto the bill if it reached his desk. However, the veto would prove unnecessary as H.R. 961 died a slow death in Senate committee. The key obstacle in the Senate was the Republican from Rhode Island, John Chafee, a moderate who thought the Clean Water Act needed updating, but not the kind of destroy-the-village-to-save-it overhaul Shuster proposed. Chafee refused to bring the bill up in the Environment and Public Works Committee, meaning that radically aggressive overreach by the GOP's extremist wing had once again doomed an antienvironmental initiative.[8]

Chafee was hardly the sole Republican opponent to the bill. Congressman Wayne Gilchrest, representing Maryland's rural, Eastern Shore–based First District, was an outspoken critic of the bill and one of thirty-four House

Republicans to vote against it. His views shed light on the GOP near the end of a transition as well as the direction the party would take. Gilchrest believed that "twenty-five years of bipartisan support for the Bay cleanup is in jeopardy," and he feared that the bill's effects "could be devastating." He diagnosed his party's problem as "a fear of information." To him, the cause was "a complete lack of knowledge about environmental issues" paired with well-funded "pressure from peer groups, interest groups, lobbyists" and, worst of all, "intimidation from other members of the House." Gilchrest was frustrated at his colleagues' unwillingness to even try to learn about the content of policy; for them it was merely an ideological contest. "If you come here [Congress] thinking you have all the information you need, and you're not going to take the time to learn about the other issues; and you have a particular doctrine that we need to reduce government, and you slash and burn your way through, you're going to undo some bad things that needed to be undone, but you're also going to undo some really good things that should stay." He also recognized that the environment was not as salient an issue as it once had been. "I went to a series of town meetings over the last couple weeks," Gilchrest said, "and almost no one talked about the environment." Gilchrest believed that "a wake-up call" was needed to maintain momentum on environmental issues. Gilchrest was especially perceptive, though the wake-up call he sought would not arrive for another two years. As 1995 wound to a close, the best Chesapeake environmentalists could hope for from the federal government was to hold the line and try to avoid a catastrophe.[9]

Chesapeake environmentalists encountered significant resistance to their policy proposals and faced aggressive efforts to weaken Bay protection at the state level as well. In Maryland, long the bastion of pro-Bay sentiment, CBF's grassroots coordinator, Susan Carter Brown, wrote in January 1996 that "CBF will work to safeguard environmental protections in Maryland from the regulatory reform fever that has swept through Congress." She laid out a bullish agenda; however, the foundation was forced to admit a year later that it had been reduced to "simply playing defense." An attempt to block William Goldsborough's appointment as one of Maryland's three representatives to the Atlantic States Marine Fisheries Commission is a case in point. Goldsborough was CBF's senior fisheries scientist and a key player in virtually every issue related to Chesapeake fisheries. He would go on to garner widespread respect for his service on the ASMFC, becoming the first person to win three separate awards from the commission, including the Hart Award, the commission's highest award. However, in 1996 a bipartisan group of seven state senators led by J. Lowell Stoltzfus, an Eastern Shore Republican, wrote to Governor Parris Glendening, a Democrat, demanding that Goldsborough be removed from the commission. The state senators "question[ed] his ability to properly assess fisheries stocks" and called him "biased towards extreme, economically damaging regulations." In addition, they warned Governor Glendening that "Mr. Goldsborough would be a political liability to you" because "his strong opponents would include watermen and packers all around the Bay."[10]

The animus toward Goldsborough was a more civil response to the same irritant that had prompted the arson in Tylerton: blue crab regulations. In the winter 1995–96 harvest season, oyster stocks continued their freefall, and while the resurgence of striped bass offered some relief, blue crabs were the major financial lifeline for commercial fishers. Unsurprisingly, blue crabs were also being dangerously overfished as more and more watermen depended on crabs to counter losses elsewhere. Fortunately, after the striped bass fishery demonstrated the value of long-term scientific data, Maryland and Virginia begun a similar monitoring program for blue crabs in 1990. Since crabs overwinter in the Bay's muddy bottom, it is possible to obtain a fairly accurate assessment of the total number of crabs in the Bay through an annual winter dredge survey. The 1995 survey, conducted by the Virginia Institute of Marine Science and Maryland's Department of Natural Resources at fifteen hundred sites across the Bay, indicated a clearly negative trend in crab populations. Goldsborough's proposal centered on creating a "sanctuary" for crabs in waters deeper than forty feet to allow fishing to continue while alleviating harvest pressure on the species. William Baker went out of his way to say that "a total moratorium on crabbing must be avoided at all costs." Governor Glendening supported moderate measures to restrict the blue crab harvest, but CBF in general and Goldsborough in particular took a significant share of the blame despite making a good-faith attempt to thread a very fine needle between supporting blue crabs as a species and supporting the communities that depended on them. As Goldsborough later recalled with a grimace, trying to save the blue crabs only "pissed everybody off at us again."[11]

Goldsborough nearly lost his position on the ASMFC because of his advocacy for the blue crabs. Fortunately for him, Goldsborough had a powerful ally in Eugene Cronin. Nearing the end of his career, Cronin was a living legend in the Chesapeake scientific community whose research expertise just happened to be on blue crabs. He was also, conveniently, a former chair of the ASMFC's Scientific Advisory Committee and a longtime ally of CBF. In one of his final contributions to Chesapeake advocacy, Cronin defended Goldsborough and the blue crab regulations in a letter to Governor Glendening. Writing that Goldsborough "has been criticised, quite unfairly, for his efforts to prevent over-harvest of blue crabs when several warning signals were noted," Cronin stated that "Mr. Goldsborough is probably the best person in Maryland outside of the State government" to nominate to the ASMFC. Persuaded in part by Cronin, Governor Glendening kept Goldsborough on the commission. It was a small but crucial victory with reverberations beyond the watershed. At the time, Goldsborough was the only representative of a conservation group in the entire ASMFC. Although Chesapeake advocates were limited to playing defense, at least in Maryland it was the kind of defense that wins championships.[12]

South of the Potomac River, Virginia had always been Robin to Maryland's Batman in Chesapeake protection, but under Governor George Allen, a Republican very much not in the Mathias/Gilchrest mold, the commonwealth took a heel turn. Allen would go on to the U.S. Senate and earn a lifetime score of 1 percent from the League of Conservation Voters, with his gubernatorial record offering a clear preview of his policy positions. Allen signaled his intentions early in his tenure by appointing Becky Norton Dunlop as Virginia's secretary of natural resources. Dunlop brought environmentalists a foreboding sense of déjà vu, recalling the dark Anne Gorsuch days. Dunlop had been a Reagan appointee at the Department of the Interior during the waning days of his administration, and like Gorsuch, she had been forced by Congress to resign. Her resignation in 1989, after just seven months on the job, had come amid allegations that she had politicized the civil service. Like Gorsuch's appointment to the EPA, Dunlop's appointments in the federal government and in Virginia were strictly political; Dunlop had no prior experience or concern about environmental issues, but she had worked for years as a conservative activist for the American Conservative Union.[13]

Dunlop did her level best to prove the maxim that history repeats itself, first as tragedy, then as farce. Her leadership of the DNR almost immediately prompted the state's Joint Legislative Audit and Review Commission to investigate the Department of Environmental Quality (DEQ). The committee's report found that after just a year under Dunlop, "DEQ's current leadership has significantly changed the emphasis, structure, and approach that the agency takes in fulfilling its statutory mandate," with negative effects on agency morale and effectiveness. It charged that Dunlop had politicized the DEQ's work, citing as evidence that 57 percent of its technical staff had responded to a survey question that they feared their jobs would be at risk if they made a decision contrary to the interests of the regulated community. Whereas Reagan eventually cut bait with Gorsuch, Allen stood by Dunlop and even chose to fan the flames by hiring the daughter of a convicted Watergate conspirator and inflammatory right-wing talk show host, G. Gordon Liddy, to a sweetheart position as deputy director of the Department of Conservation and Recreation, a position that Allen had eliminated two years earlier when slashing the department's budget. A spokesperson for Dunlop made no effort to conceal the naked political calculus; the decision was justified because Allen needed someone in the agency with "a loyalty to the director and to the secretary and to the governor." Dunlop survived Allen's term, but her scorched-earth tactics were so extreme that Allen's successor, fellow conservative Republican James Gilmore, deliberately leaked a memo during his campaign promising to fire Dunlop if elected. She would resign on her own and then promptly join the Heritage Foundation as its Ronald Reagan Distinguished Fellow.[14]

In a sign of the shifts in public opinion, Allen remained a popular governor despite his antienvironmentalism. Virginia drew national attention for its poor environmental record when the EPA threatened to cut off the commonwealth's authority to issue permits under the Clean Air Act. The crux of the problem was Virginia's strict citizen-standing rules, which prohibited an individual from challenging an air-pollution permit unless they could prove immediate and major financial loss. The EPA threatened to prevent Virginia from receiving $500 million in federal transportation appropriations until the commonwealth brought its air-pollution permitting process into compliance with federal law. Allen's response was political grandstanding and countersuits. His legal strategy was more political rhetoric than thoughtful jurisprudence, presaging coming changes to American conservatives' political style. He made a classic states'-rights argument that the "reserved powers clause" of the Tenth Amendment to the U.S. Constitution made the EPA's authority to intervene in the state program under the Clean Air Act unconstitutional.

Allen attempted to turn the EPA battle into a conservative cause célèbre—he made a highly publicized march from the state capitol to the federal courthouse in Richmond to deliver his lawsuit—but while his position was ideologically appealing to the Right, it held little legal water, at least at that juncture in history. Allen lost his battle with the EPA over the Clean Air Act in 1997 after taking the case all the way to the Supreme Court. Although Virginia ultimately passed a law to be in compliance, Allen's actions were an early indicator of the direction conservative lawmakers and activists would take to try to undermine and undo policies they opposed through the courts.[15]

For the Chesapeake, Allen's tenure had many obvious negative consequences, but some of the worst did not come to light until after he left office. Investigative reporting in 1999 by Ron Nixon at the *Roanoke Times* revealed a cover-up involving Virginia's program for monitoring toxic pollution. Under Allen and Dunlop, the DEQ had refused to allow anyone access to the Virginia Toxics Database, claiming that it had been destroyed after Allen eliminated the program in 1994. Even though Virginia's monitoring program ceased, the database still contained valuable information about historical levels of toxins in Virginia's waters, and both scientific researchers and natural resource management officials, including the U.S. EPA, requested access to the files. Nixon reported that those seeking information "were told that the database was dead. Some were told that the data were in an inaccessible format and that there was no one to obtain access to the data. Others were told that the data were too old, damaged, incomplete or of poor quality." State officials were lying. Nixon found that in reality "the database spent more than five years locked in a safe on the second floor of an office at the agency's building in Richmond." It was not until a whistle-blower within the DEQ discovered that the database was mostly intact and notified VIMS and the EPA that the story came to light. In an editorial following Nixon's article, his newspaper called the incident "an outrageous example of public officials manipulating a state agency and public documents for ideological purposes." Once exposed, the DEQ made the data publicly available, but this episode, like many others during Allen's tenure, underscores the extreme rightward drift of conservative Republicans and the resulting difficulty environmentalists faced in Virginia and nationally when trying to advance their causes.[16]

Finally, the situation in Pennsylvania mirrored that of Maryland, with CBF merely "holding the line" as the wave of small-government zeal crashed over the watershed. At least in Pennsylvania there was no overt hostility from Republican governor Tom Ridge. In fact, he appointed CBF's Pennsylvania office director, Jolene Chinchilli, to the state Department of Environmental Protection's Citizen Advisory Council. Chinchilli was the only Chesapeake-connected member of the council, making hers a critical, if lonely, voice. Despite more cordial relations with CBF, there were not many bright spots in Pennsylvania. In 1995, "the most significant environmental legislation passed" was a law reorganizing the Department of Environmental Resources into two new cabinet-level agencies, the Department of Environmental Protection and the Department of Conservation and Natural Resources. The stated goal of the change was—ominously, to environmentalists—to "make the state more user friendly to those it regulates." The following year, Ridge's administration made good on that promise and dealt a major blow to the Chesapeake by removing wetlands protection. CBF had identified preserving the watershed's remaining wetlands and rebuilding new wetlands as a top organizational priority, and the Harrisburg office spent most of 1995 opposing a bill that, if passed, would redefine half the state's wetlands out of existence. Despite vigorous opposition in the General Assembly, CBF's efforts came to naught when the Department of Environmental Protection issued General Permit 15, which allowed developers to fill up to half an acre of wetlands per lot. CBF legal challenges to this permit failed, leaving William Baker to complain that "large developers have been behind the lobbying blitz for this action. The new law will allow vast wetland areas, subdivided into half-acre lots, to be filled. No cumulative assessment of the total number of acres destroyed has even been made in spite of CBF appeals to do so." It was a blow to the organization, and to the broader movement, to lose so convincingly on such a critical issue affecting the Bay's health. Across the watershed, from the state capitals to Congress, the momentum to Save the Bay had stalled.[17]

PFIESTERIA HYSTERIA

What shook Chesapeake environmentalism out of its stupor sounds like something ripped from the pages of a Michael Crichton novel. In the summer of 1997 an invisible, shape-shifting menace capable of killing its victims by melting the flesh off their bones stalked the Bay's waters. It was the toxic dinoflagellate *Pfiesteria piscicida*, whose scientific name literally means "fish killer." Although it sounds fictional, *Pfiesteria* was quite real: dinoflagellates are microorganisms capable of transforming from plant-like to animal-like morphs during different stages of their life cycle; in certain stages, the organisms release powerful biotoxins that can dissolve the living tissue of affected fish. *Pfiesteria* was such a strange and mysterious threat that it spawned several sensationalist works, including the nonfiction book *And the Waters Turned to Blood*, in which the author warns *Pfiesteria* "may become the ultimate biological threat"; the novels *Sea Change*, in which *Pfiesteria* threatens to obliterate Seattle, and *The Swarm*, in which it is part of the ocean's collective effort to destroy humanity; and finally, on the big screen, Barry Levinson's eco-horror flick *The Bay*, in which *Pfiesteria* wipes out a small town in Maryland. Although the microorganisms were not ultimately a threat to the Bay, much less humanity, they scared a lot of people into paying attention to the Chesapeake again. According to the CBF attorney Tom Grasso, "*Pfiesteria* has been the wake up call to a much larger nutrient pollution problem that we need to address." Unlike nutrient pollution, *Pfiesteria* has not proven to be a chronic problem for the Bay. Indeed, there is not much clear evidence that *Pfiesteria* caused widespread fish kills, however, at the time, it became a frightening symbol of the Bay's ecosystem being dangerously out of balance. Although the microbe was not a major threat, the media mania surrounding it gave environmentalists an opening to launch a new round of policies that broke through partisan gridlock, at least in the watershed states.[18]

In his excellent article "How Did a Media Storm Get Started? The Frenzy over *Pfiesteria*," Michael Fincham describes how unexplained fish kills in some remote Eastern Shore tributaries became national news. In May 1997 Brad Bell, a TV reporter for WJLA, the local ABC affiliate, went to Shelltown, on the rural backwaters of the Pocomoke River (see map 2), to investigate reports of dead and dying fish covered with mysterious sores and lesions. His story struck a chord with viewers, and despite (or perhaps because of) reassurances from Maryland state officials, Bell decided to return to the Pocomoke River to collect a sample and deliver it to JoAnn Burkholder, who had codiscovered *Pfiesteria piscicida* several years earlier and linked it to massive fish kills in North Carolina estuaries. As Fincham relays the story, "In an act of pure enterprise reporting, Bell headed back to the Pocomoke, filled

Menhaden with lesions associated with *Pfiesteria piscicida*. (Courtesy of Center for Applied Aquatic Ecology, North Carolina State University)

an empty Evian bottle with water from a fish pound, then drove down Interstate 95, and delivered his sample personally to Burkholder's lab in North Carolina. Peering through a light microscope, Burkholder told Bell that the water held cells that looked like Pfiesteria. A detailed analysis, including fish bioassays and electron-scanning microscopes, would take a couple weeks, but Bell was not waiting. His report aired that evening." Bell's report blew the lid off the story, and despite their best efforts at transparency, Maryland state officials were unable to provide a clear explanation of the fish kills, prompting further investigation, which escalated into a coverage war between the *Baltimore Sun* and the *Washington Post*. It was an enticing cocktail of local news, scientific mystery, and environmental calamity, with a garnish of political blundering. Although the *Sun* and the *Post* were competing for local readership—the *Post* published 130 Pfiesteria stories; the *Sun*, 170—the *Post*'s stature as a paper of record made the issue national news, especially as more and more reports of people experiencing neurological symptoms connected to the fish kills emerged. The capstone to the calamity came over Labor Day weekend in 1997, when Governor Glendening closed the lower Pocomoke out of fear of the potential human health hazards from *Pfiesteria*. It was a terrible blow for the Pocomoke's communities in particular and for the Chesapeake as a whole. The Maryland Department of Agriculture estimated that the total economic impact on commercial and recreational fisheries, along with lost revenue from tourism, came to $127 million. It was in many ways the Bay's biggest disaster, or at least its most publicized, since Tropical Storm Agnes had devastated the region twenty-five years earlier.[19]

Like Agnes, *Pfiesteria* marked a turning point in the Bay's environmental history. This was not so much because of the severity of the damage it caused, because the summer of 1997 remains the only major *Pfiesteria*-associated fish kill in the Chesapeake, but because reports of a fish-killing, human-sickening menace shone a spotlight on the very real problems posed by agricultural pollution. A decade prior, the 1987 Chesapeake Bay Agreement had set a goal to reduce nutrient pollution by 40 percent before the year 2000. In the subsequent years, environmentalists had made little headway on that goal in part because of the strong political headwinds they faced, but also because for many people nutrient pollution simply did not sound like a problem. Wetlands loss, shoreline erosion, deforestation, fisheries' collapse, and the like, were all threats people could see and easily understand. Thanks to *Pfiesteria*, CBF staff were "pleased with the growth of public understanding of pollution from nitrogen and phosphorus" and planned to target agricultural pollution, highlighting the link between manure and *Pfiesteria*. William Baker called *Pfiesteria* "a metaphor for all that is wrong with the Bay" and "nature's clarion call to action, a rude awakening for all who have become complacent about the need to save Bay." Even as the crisis was still unfolding, Baker and others saw a newly awakened public as "the potential silver lining in the dark Pfiesteria cloud."[20]

Unsurprisingly, Maryland was the first state to act, both as the epicenter of the *Pfiesteria* outbreak and as the traditional leader on Chesapeake issues. On September 15, 1997, Governor Glendening appointed a blue-ribbon "action commission" to make recommendations for addressing the *Pfiesteria* problem. He gave the commission a November 1st deadline to report back with policy proposals for the next General Assembly legislative session. Glendening stacked the commission with big names in local environmental politics, including Baker and the prominent proenvironment legislators Bernie

Fowler and Brian Frosh, and tapped the venerable Harry Hughes, former governor and Chesapeake champion, to chair the commission. Glendening included two state legislators from the Eastern Shore, though these would be the only two to vote against adopting the commission's recommendations. Given this makeup, it came as no shock when the committee recommended nutrient controls on farm runoff. While the committee's final report emphasized that "all Marylanders are in this together" and "no one person or industry is responsible for causing this problem or for polluting the Bay, and altering one type of behavior or practice is not going to make these problems go away," the bulk of the report focused on issues related to agricultural nutrient pollution.[21]

Undergirding the commission's findings and recommendations were the conclusions from a group of scientists convened at Chairman Hughes's request to provide technical advice to the commission. What became known as the "Cambridge Consensus report" explicitly linked *Pfiesteria* outbreaks to nutrient pollution. Although it was not a peer-reviewed study, the weight of scientific consensus, media hype, and a strategic choice of blue ribbon commissioners yielded a somewhat surprising result for the Bay. Six months after they submitted their recommendations, the Maryland General Assembly passed the Water Quality Improvement Act of 1998. Although the bill contained typical drawbacks—light penalties for violators, a multiyear grace period before implementation, key exemptions—it was still a major step toward controlling nutrient pollution, and it likely would not have been possible without the fear of *Pfiesteria*. As the *Baltimore Sun* reported at the time, "Even after being softened, the bill retains mandatory controls on farm nutrient runoff—a provision environmentalists know they would never have achieved without last year's *Pfiesteria* outbreaks in three Maryland waterways." Even so, CBF's Tom Grasso said, "We're pleased that limits have finally been set, but we are disappointed that the bill as passed will not address the immediate nutrient pollution problem." Nevertheless, three years later one scholarly analysis concluded, "It is widely believed that the requirements of this legislation make it the most comprehensive agricultural nutrient management law in the nation."[22]

South of the Potomac, Virginia's seafood industry suffered a blow even as most of the fish kills took place in Maryland waters. There are no cumulative figures for Virginia, but the *Richmond Times-Dispatch* reported that in some areas sales of seafood declined by nearly 50 percent as "news of its [*Pfiesteria*'s] scattered outbreaks around the Chesapeake Bay this summer and fall has crippled the state's half-billion-dollar seafood economy." A 2003 study of *Pfiesteria* fish kills and their economic impact estimated that the economic loss stemming from a *Pfiesteria* fish kill could range from $37 million to $72 million in the month following the kill, largely because of decreased consumer demand. Given Maryland's losses, the anecdotal evidence from Virginia, and how highly publicized the Chesapeake's *Pfiesteria* fish kills were throughout the summer months, it seems safe to conclude that the total economic damage to Virginia exceeded $100 million. This economic context helps explains why Governor Allen, normally loathe to spend money on environmental causes, was willing to authorize $2.3 million to take action against *Pfiesteria* in the waning days of his administration. Bolder measures would have to wait for the tenure of his successor, James Gilmore.[23]

As noted above, Gilmore explicitly sought to distance himself from Allen's environmental record. With his support, the Virginia General Assembly passed a two-year budget early in 1998 containing a then record $57 million in dedicated Chesapeake Bay spending to target nutrient pollution. Funds included upgrades to wastewater treatment facilities, support for planting riparian buffers along waterways, and financial incentives to farmers to reduce agricultural nutrient pollution. It was the most comprehensive attempt ever by the state of Virginia to reduce nutrient pollution, though there were some loopholes, the most concerning of which exempted poultry-growing operations from regulations that applied to hog, beef, and dairy producers. Representatives of the poultry industry were able to delay, but not defeat, the eventual inclusion of their operations under state regulations, especially as CBF led a steady drumbeat connecting poultry waste with *Pfiesteria*. In a typical example, a CBF publication stated that "most Virginians will be surprised to learn that poultry houses are not even covered under current state environmental regulations! This despite the fact that 90 percent of poultry operations in the state reside in the Bay drainage basin, including the Delmarva peninsula, where they are suspected to contribute to the outbreak of Pfiesteria." The *Pfiesteria* threat had reenergized public concern about the environment. One state legislator reported receiving hundreds of letters and phone calls about the poultry waste bill, compared with just a handful about other high-profile legislation. In early 1999 the General Assembly voted unanimously to close the poultry loophole. Joseph Maroon, director of CBF's Virginia office, bragged that "CBF staff and its members made poultry manure regulation an issue Virginia legislators could not ignore." Although perhaps overstating the case, Maroon deserved a victory lap. Amazingly, a state that only two years earlier had seemed determined to be as recalcitrant as possible was now vying with Maryland for the lead in Chesapeake protection.[24]

Pfiesteria was less of a catalyst in Pennsylvania than it was in Maryland and Virginia, given that the state had not suffered the same economic harm as its southern counterparts. Nevertheless, after 1997 environmentalists attacked the problem of agricultural nutrient pollution with renewed vigor in Pennsylvania. Building on the foundation laid by the Nutrient Management Act, in 1998 the Pennsylvania Department of Environmental Protection agreed to issue new interim guidelines for industrial-scale livestock operations known as concentrated animal feeding operations, or CAFOs. The following year, the state adopted final regulations that required CAFOs to have manure storage systems and state-approved nutrient management plans. Environmentalists' biggest victory in Pennsylvania came at the end of 1999, when the state committed $645 million over the next five years to a new, statewide environmental protection and restoration program called Growing Greener. While not Chesapeake-specific, the measure was still a huge asset for protecting tributaries, forests, and wetlands in the Pennsylvania portion of the watershed.[25]

Although the *Pfiesteria*-related fish kills have thus far proved to be a one-off event, they made a lasting impact on the region. A 2000 study found that 85 percent of residents from the watershed states believed that money being spent by their state on *Pfiesteria* was a good investment. Crucially, the study authors found that "runoff from livestock and agricultural operations, along with pollution from factories and cities, were the most often cited responses to queries about the factor contributing to Pfiesteria." Clearly, the crisis had heightened public awareness of the severe impact of pollution from farms on the Bay's health and increased support for governmental action, which helped drive policy change at the state level throughout the watershed. Considering how much difficulty Chesapeake environmentalists faced before *Pfiesteria*, the frightful summer of 1997 ranks with Tropical Storm Agnes in 1972 and the Chesapeake Bay Agreement in 1983 as a crucial turning point in the Bay's environmental history.[26]

A NEW WAY TO SAVE THE BAY

The final and most lasting impact of the *Pfiesteria* scare was that it helped kick off a new round of regional planning for protecting and enhancing the Chesapeake Bay. This culminated in a new agreement, Chesapeake 2000, which was supposed to guide the Bay's restoration into the twenty-first century, with numerous, specific restoration goals designed to hold leaders accountable. Chesapeake 2000 was hailed by national media, such as the *New York Times*, "as an environmental model for the nation." Addressing a broad range of issues from habitat restoration to environmental education, many of the agreement's provisions were adopted explicitly in response to environmentalists' critiques of the shortcomings of prior approaches to the Chesapeake Bay Program. It was bold, it was ambitious, and it was a total failure. Several factors contributed to Chesapeake 2000's shortcomings, but the biggest challenge was an inability to get a significant commitment from the federal government to fund and enforce the policies necessary to sustain the Bay. There were warning signs before the ink dried on the agreement. For instance, notably absent from the flurry of post-*Pfiesteria* legislation were any major Chesapeake-specific measures from the EPA, Congress, or the White House. Furthermore, Congress dragged its feet when it came to reauthorizing and reallocating funds for the EPA to continue the Chesapeake Bay Program. This trend continued in the early years of the twenty-first century, with the absence of federal involvement growing more glaring as environmentalists scored bigger and bigger victories at the state level. The paradox of Chesapeake 2000 is that it ended up being extremely important, just not for the reasons its architects had hoped.[27]

The person perhaps most responsible for the Chesapeake 2000 agreement was Ann Swanson, who was the principal drafter of the document. She joked that "we really were swept in to the whole Y2K thing. Everybody was doing stuff associated with 2000 and we thought we would too." More seriously, she acknowledged that the previous, 1987 agreement had failed miserably to reach the goal of reducing nutrients by 40 percent and said that it was time to renew the effort. As Swanson saw it, the pieces were all present for revisiting the regional strategy to restore the Chesapeake; however, it was not until after *Pfiesteria* that Swanson was directed to begin working on a new agreement. Swanson identified *Pfiesteria* as the "catalyst" that set the process in motion. As she later explained, "The politicians had to show they were responding. They had to have an action, and it couldn't just be scientists studying it, right? Because people had the sense that there was polluted water and the polluted water was allowing this brain-munching organism to thrive. And so they signed a far reaching agreement and passed progressive legislation aimed at curbing pollution. All the cards aligned. . . . So it's all a matter of timing, when the right information and conditions come together. We had the facts, now we had the action."[28]

Despite the cards being aligned, drafting the agreement was a brutal, two-year process of negotiations, horse trading, and political brinkmanship. Yet, when Chesapeake 2000 emerged, it elicited real hope because it contained so many specific provisions written with input from environmentalists and other stakeholders. This was in sharp contrast to the two prior agreements, which were brief, vague, and written without much public input. Swanson's perspective is invaluable because she was CBF's grassroots coordinator in the 1980s before becoming the Chesapeake Bay Commission's executive director, a post she would hold for thirty-five years. She recalled that "CBF was not involved in the negotiation of the '87 Bay Agreement. Absolutely peripherally. And I remember being really troubled by that." Swanson deserves tremendous credit for opening up the process of drafting Chesapeake 2000, but in the intervening years CBF had made steps to ensure that it would have a seat at the table. Following the 1992 amendments to the 1987 Chesapeake Bay Agreement, CBF staff and trustees began working on an internal, long-term strategic plan. One of their key insights from their strategic planning discussions was that despite not reaching its target, "the Bay Program's 40 percent nutrient reduction goal has been a powerful motivator, a specific measure of progress that holds officials' 'feet to the fire.' Other programs, lacking such specific goals, are less effective." In addition to this nutrient goal, CBF's long-term plan identified several specific targets related to wetlands, submerged aquatic vegetation, and water clarity and sought to identify more key targets. Crucially, in the final version of the plan—approved by the board of trustees in January 1994—CBF resolved, "We will measure success by specific indicators of the Bay's health and strive to convince others to adopt *the same indicators*." Intriguingly, the name of this long-term plan was Chesapeake 2000.[29]

As a result of these strategic planning discussions, CBF eventually developed nine indicators, which it published in 1995 with ten-year benchmarks to be reached by 2005. These indicators were water quality, dissolved oxygen, toxic pollution, submerged aquatic vegetation, migratory fish, oysters, riparian buffers, wetlands, and resource lands. All but water clarity and dissolved oxygen became part of Chesapeake 2000 and have been tremendously influential in shaping the Chesapeake restoration effort over the past quarter century. While the goals for the most part went unmet, the underlying logic behind them—to have specific targets to hold political leaders accountable—has proven quite sound and is a key reason why the Bay's water quality and living resources have improved, if not to the degree environmentalists had hoped. One final aspect of these indicator goals merits attention. In addition to being key components of Chesapeake 2000, they would also form the basis for CBF's first *State of the Bay* report, published in 1998. The report was highly influential and it has been variously copied, criticized, and celebrated since its inception. It was ultimately a highly effective public relations tool even if foundation insiders privately conceded that it was more impressionistic than scientific. The 1998 report gave the Bay a score of 27 out of 100 (with a saved Bay defined as one that would score 70 out of 100). In the publication, CBF acknowledged that "this simple number describes the health of a nearly unimaginably complex system woven from the threads of the living and the physical worlds." One of the unintended consequences of reducing the Bay's health to a simple number is that these numbers have consistently portrayed the Bay's recovery as "failing." This has been at times useful for generating support for Bay-friendly policies—after all, a score of 27 is shockingly bad—but this approach has also obscured the restoration movement's successes, and most crucially, it has hidden the complexity and difficulty inherent in improving the Chesapeake ecosystem. Thus, major accomplishments are recast as failures to achieve goals that were unrealistic in the first place. CBF's decision to adopt specific indicators and reduce the Bay's health to a single number was a double-edged sword; while it was useful in spurring progress toward important goals, it also oversimplified the complexity of the restoration movement and contributed to a dynamic in which perceived failures overshadow tangible accomplishments.[30]

CBF had more immediate problems to worry about in 1995 than how subsequent interpretations of their indicator goals would influence public opinion about the Bay's recovery. As made clear above, the mid-1990s political climate was not favorable for debuting an ambitious and expensive plan to save the Bay. CBF's membership declined by 13 percent from 1994 to 1996, after twenty straight years of growth. Worse still, the organization found itself stretched so thin financially that it decided not to publish an annual report in 1996. Apologizing to its members, the foundation issued a statement that "savings incurred will enable us to issue other needed materials and publications during the coming months." All this came while CBF was trying to persuade philanthropists and politicians to adopt its restoration goals and spend millions more to achieve them. It is inconceivable that CBF's proposals would have been adopted so swiftly or enthusiastically, if at all, without *Pfiesteria*. While correlation is not causation, the spike in CBF's membership after 1997 is striking. The Bay was in a crisis, and fortunately, the region's oldest environmental group had a plan ready to go. In a savvy update from 1995, CBF explicitly linked its own numeric restoration goals to *Pfiesteria*, human health, and the region's economy. Citing oysters as a prominent example, William Baker called on the region's political leaders to follow CBF's lead and invest in restoring oyster reefs "to produce the billions and billions of oysters needed to Save the Bay, help prevent Pfiesteria piscicida from making humans ill, and restore thousands of jobs to hardworking, local watermen." Following immediately on the heels of the economic harm and public health scare from *Pfiesteria*, this pitch was especially persuasive, and the idea of ecosystem restoration—planting more trees and underwater grasses, rebuilding wetlands and riparian buffers, restocking barren oyster reefs—became central to the planning for Chesapeake 2000.[31]

The process of creating a new Bay agreement came to a conclusion on June 26, 2000, when the signatories to the previous two agreements—Maryland, Virginia, Pennsylvania, the District of Columbia, the U.S. EPA, and the Chesapeake Bay Commission—renewed their commitment "to the restoration and protection of the ecological integrity, productivity and beneficial uses of the Chesapeake Bay system." The thirteen-page document contained dozens of specific, deadline-driven goals the signatories agreed to reach by 2010. In a bittersweet inclusion, Chesapeake 2000 adopted the 40 percent nutrient reduction goal that its predecessor agreement had miserably failed to meet. Some of the most noteworthy restoration goals were a tenfold increase in the number of oysters in the Bay and replanting more than 100,000

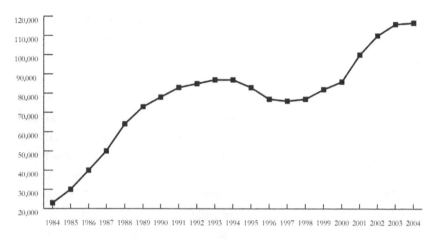

Chesapeake Bay Foundation membership, 1984–2004. (*Source:* Annual Reports, CBF internal records, Philip Merrill Environmental Center, Annapolis, MD)

acres of underwater grass beds, 2,000 miles of forested riparian buffers, and 25,000 acres of wetlands; the agreement also set goals to reopen dammed streams for migratory fish and set harvest targets to create a sustainable blue crab fishery. All these elements were derived from the indicator goals CBF had developed in 1994. Additionally, the signatories agreed to reduce the rate of "harmful sprawl development" by 30 percent and permanently preserve 20 percent of the land area from development. These goals were especially contentious and hard-won. According to one senior Virginia official, at first "the proposal to reduce the rate of conversion of forest and agricultural lands to development by at least 30 percent by 2010 was impossible for the Gilmore administration to accept." Governor Gilmore was the only member of the Chesapeake Executive Council to oppose the goal, although he relented after reaching a compromise to make the reduction 30 percent watershed-wide instead of 30 percent in each state and to apply only to "harmful" sprawl, leaving the term conveniently ill-defined. Nevertheless, these measures made Chesapeake 2000 "the nation's first regional attempt to contain urban sprawl and permanently protect thousands of acres of farm and forestland." Less heralded but still far-reaching commitments included "to provide a meaningful Bay or stream outdoor experience for every school student in the watershed before graduation from high school" and to expand public access to the Bay by 30 percent. These and similar measures provided a crucial means for continuing to deepen public engagement and concern about the Bay.[32]

All told, Chesapeake 2000 was a remarkably thorough and thoughtful effort to make a plan to protect and restore the Chesapeake. It was not perfect. It included plenty of compromises, and probusiness politicians like Governors Gilmore and Ridge certainly got concessions, but environmentalists got a lot of what they wanted too. Tom Horton, who by this point had been covering the Chesapeake environmental beat for more than twenty years and was not shy about pointing out flaws where he saw them, criticized the agreement for various shortcomings, primarily that it continued to be a voluntary pact and that it was vague in places and contained omissions in some key areas. "Still," he acknowledged, "it's amazing everyone agreed to this much." Likewise, William Baker lamented "glaring" omissions about dumping dredged materials and air pollution; nevertheless, he concluded that "this new agreement is moving beyond simply stopping the Bay's decline. It addresses actual restoration. . . . CBF members, trustees, and staff can take great pride in this evolution as it is consistent with the agenda we have advocated. In fact, many of the agreement's components are close, if not identical, to our own set of goals." This was a triumph. In just a few short years the foundation had gone from the political wilderness to having its plan for saving the Bay adopted as the guiding framework for the whole Chesapeake Bay Program. If only it had worked.[33]

WHY CHESAPEAKE 2000 FAILED

For all Chesapeake 2000's potential to transform the estuary and its watershed, it mostly failed to meet its goals. The two biggest problems were that the agreement was a voluntary project lacking any enforcement mechanisms and that no one had any idea how much it would cost to implement its provisions or a plan to finance them. The failure to implement Chesapeake 2000's provisions would highlight the need to finally address these problems, which required leadership from the federal government. As we will see, funding and enforcement mechanisms have not been silver bullets for the Bay's restoration, but recent improvements in the Bay's health would not have been possible without them or without the U.S. government's willingness, at long last, to treat the Bay like the national treasure it is. These final pieces of the Bay's sustainability puzzle would only be put into place after Chesapeake 2000's spectacular failure. Explaining the agreement's failure requires incorporating national and global issues into the Chesapeake's environmental history. World-historical events as far-reaching as the war on terror starved environmental projects of attention and money, while at the national level the George W. Bush administration was more hostile to environmental protection than any of its predecessors. More locally, the Chesapeake Bay Foundation made a strategic error by focusing most of its energy on implementing Chesapeake 2000 at the state level. The confluence of these factors meant that the federal government would do little more than the bare minimum for its part of the Chesapeake Bay Program, dragging down the entire restoration effort. The inescapable conclusion is that in the early years of the twenty-first century environmentalists missed an opportunity to have a much healthier Bay at a much smaller cost to the public.

The first attempt to estimate the cost of implementing Chesapeake 2000 came from the Chesapeake Bay Foundation, which calculated that $8.5 billion spent strategically over the ten-year window would achieve the agreement's main goals. CBF's estimate focused on key elements of the agreement that closely matched the foundation's own goals: upgrading sewage treatment plants and improving storm water infrastructure, planting forested riparian buffers, preserving land, funding agricultural nutrient management plans, and restoring wetlands and oyster reefs. CBF proposed that the federal government provide $4 billion of the total, a figure the three main watershed states would collectively match, with the remainder coming from local and private sources. A key frame of reference for CBF's estimate was the Comprehensive Everglades Restoration Plan, or CERP. Around the same time that the watershed states signed Chesapeake 2000, the federal government and stakeholders in Florida reached an agreement on a $7.8 billion, thirty-year plan to restore the Everglades. Observers touted CERP as the "largest and most expensive ecological restoration effort ever undertaken" and "the most ambitious effort in recent U.S. history to reform a water management infrastructure." CBF took note of the largess the Everglades received. When announcing CBF's plan, William Baker specifically referenced "the precedent-setting decision to spend $7.8 billion on the Florida Everglades," which he argued "makes the $8.5 billion price tag for [the more populous Chesapeake Bay] watershed, a multi-state resource, all the more reasonable." The near fifty-fifty cost sharing in CBF's plan also drew its inspiration from the Everglades plan.[34]

CBF presented its proposal to the Chesapeake Bay Commission in May 2001. This ruffled some feathers as the commission was in the midst of developing its own, substantially higher cost estimate ($18.7 billion); nevertheless, the commission's staff and members acknowledged CBF's leadership on the Chesapeake 2000 goals. The commission, "supported by EPA's technical expertise and CBF policy staff," chaired an informal working group affectionately dubbed the "Lego Workgroup" tasked with identifying the most important building blocks and arranging them in a way to maximize the chances of successfully achieving Chesapeake 2000's targets. Initially the prospects of the Lego Workgroup seemed bright. For example, in June 2001 Senator Paul Sarbanes (D-MD) introduced the Chesapeake Bay Watershed Nutrient Removal Assistance Act, which proposed to allocate $660 million over five years to upgrade the region's sewage treatment plants. This was a moonshot bill that would have met a third of the federal government's annual contribution to the Bay's restoration under CBF's proposal had it been passed. It is impossible to say whether the bill would ultimately have made it through Congress, but it had bipartisan support from all six senators from the watershed states, including a former nemesis turned junior senator from Virginia, George Allen. Anecdotally, there seemed to be plenty of support for the measure. In Pennsylvania, a state where Chesapeake-specific measures typically had a hard time gaining traction, CBF found "some of the most widespread grassroots support from Pennsylvanians on any federal initiative CBF has ever undertaken." However, the bill never had a chance because the terrorist attacks of September 11, 2001, completely eradicated environmental issues from the national agenda.[35]

The historical scholarship analyzing how the global war on terror affected and interacted with U.S. politics and environmental issues is still in its infancy, and while broader conclusions are well beyond the scope of this work, several preliminary observations pertaining to Chesapeake environmentalism may be of use for framing future research questions. As others have noted, "the enormous budgetary drain of the wars in Iraq and Afghanistan left little funding for environmental programs and innovation," while "the

George W. Bush doctrine of preemption and good versus evil in a neverending War on Terror obscured his less popular antienvironmental stance and left little political breathing room for any oppositional movement. Indeed, a major casualty was environmental issues, which received little attention during the campaign and a low ranking from voters." In the aftermath of 9/11, it would have been hard to justify devoting time, energy, and resources to environmental issues, even if an administration had been inclined to do so. The Bush administration was not so inclined. As one senior CBF staff member lamented, "Because of 9/11 we lost a lot of momentum in the first couple of years and then the second Bush Administration managed to keep it stifled." The early years of the twenty-first century were in some ways an even bleaker time for environmentalists than the mid-1990s. In early 2004 William Baker questioned whether "our elected officials have simply lost the political will to Save the Bay" and lamented that because the "effort to Save the Bay has been stalled for over three years . . . the failure to move forward has actually resulted in a decline in water quality." Although there had been fraught moments between CBF and politicians before, they reached a new low under the Bush administration. By Bush's second term, Baker was breaking with precedent by calling out the president and his officials by name, explicitly blaming them for Chesapeake 2000's failure. For instance, in 2007 he wrote, "The bottom line is this: if the Bush administration had met the states even halfway, we would now be within striking distance of the 2010 goal. The administration's failure to adhere to the terms of the Chesapeake 2000 Agreement—an agreement to which they were and are a full partner—is inexcusable." This was worse than Reagan, this was worse than Newt Gingrich, this was an all-time low point in the federal government's commitment to restore and preserve the Chesapeake Bay.[36]

The situation was little better at the state level, at least initially. The Chesapeake Bay Commission, which had a leading role to play in the implementation of Chesapeake 2000 across the watershed, reported gloomily that "having shepherded *Chesapeake 2000* to its successful adoption during more financially solvent times," it now faced "daunting fiscal challenges" as the regional and national economy absorbed "staggering financial blows." With state budgets deeply in the red, the only realistic way to pay for implementing Chesapeake 2000 was through federal funding, and spending money on environmental restoration simply was not a priority for the Bush administration.

While the Bush administration deserves blame for abdicating its responsibilities under Chesapeake 2000, funding the Chesapeake Bay restoration had not been a priority for previous administrations either. In 2003 the Chesapeake Bay Commission completed a detailed economic analysis of the Bay's restoration and found that historically the federal government had contributed only about 18 percent of the money spent on the Bay. This confirmed what many observers already knew—Maryland was spending significantly more on the Bay than Virginia and Pennsylvania—but the numbers were still shocking. Not only had Maryland committed more to the Bay's recovery than Virginia and Pennsylvania combined, they didn't even come close. Maryland contributed nearly 60 percent of the state-level spending on the Chesapeake Bay. Even with Maryland's generous support for the Bay, the Chesapeake Bay Commission estimated that the state had only committed to spend about 55 percent of what it would cost to implement its share of the Chesapeake 2000 goals. Thus, by 2003, with little meaningful progress toward implementing the agreement's provisions, environmentalists privately acknowledged that Chesapeake 2000 was a dead letter, even as they continued to publicly call for leaders to fulfill their promises toward the Bay.[37]

CHANGES AT CBF

These national—indeed international—events would have thrown the best-laid plans awry. Unfortunately, there was also a dynamic unique to Chesapeake environmentalism that further hampered the movement. The region's leading environmental force, the Chesapeake Bay Foundation, made a tactical error. In the strategic planning process that eventually produced many of Chesapeake 2000's goals, CBF assumed that "it is the states whose actions most directly affect the Bay's health, and it is at that level we will work primarily to influence government." This proved to be a huge blunder and a misreading of the restoration movement's history, but it was in keeping with previous efforts. A longtime CBF staffer acknowledged that despite the occasional high-profile bill, for most of the group's history "we weren't playing at the federal level all that much." Although Senator Charles Mathias's role in the Bay recovery has been overstated at times, there is no doubt that he was the Bay's biggest champion at the federal level. His retirement, combined with national shifts in the Republican Party away from environmental protection, left a massive vacuum in Washington, DC. The mid-1990s Contract with America-style opposition to environmental protection quickly became the new norm for Republicans. Chesapeake environmentalists, and CBF in particular, were slow to realize this, in part because they continued to find bipartisan support for the Chesapeake at the state level. More importantly, the focus on the states obscured the fact that federal commitment was essential to the very existence of the Chesapeake Bay Program. Without the U.S. EPA's Chesapeake Bay study (and the Reagan administration's ham-fisted response to it), there would not have been a regional effort to restore the Bay, and without the threat of potential federal involvement there would have been a near-zero chance that Pennsylvania would voluntarily sign the first Chesapeake Bay Agreement.[38]

By the middle of the first decade of the twenty-first century, CBF realized that it would need to change its tactics. One insider said about Chesapeake 2000 that "we gave it a real run for our money," but the foundation's attitude was that "we'll try to do it cooperatively over ten years, and if not, we'll sue." In other words, CBF had gotten the states to agree to its own goals, and it planned to work collaboratively with them to implement those goals. But when that did not work, the foundation transitioned to a more litigious approach that relied on using the courts to force the EPA and the states to get the Bay off the impaired-waters list as required by the Clean Water Act. In late 2003 CBF began to quietly assess what it would take to craft a legal strategy to enforce the Bay's cleanup, and in June 2004 the foundation went public with its Chesapeake Bay Watershed Litigation Project. CBF had employed lawyers on its staff since the 1970s and had participated in major lawsuits since organizing the legal challenge to the Calvert Cliffs nuclear power plant in the late 1960s, but the Chesapeake Bay Watershed Litigation Project was a new approach. As William Baker explained, the project "marks the beginning of a more proactive and strategic pursuit of legal action as a tool for Bay restoration." The person CBF picked to head its new litigation project was Jon Mueller, a veteran attorney at the Department of Justice who had been doing environmental enforcement work. Mueller had been at the DOJ for seventeen years, and what excited him about the CBF position was "the opportunity to have a little more role in the development of policy." Mueller's hire would be crucial because he would go on to lead the charge to force the U.S. EPA's hand to implement provisions of the Clean Water Act across the Bay's watershed.[39]

Mueller's hire wasn't the only key strategic change for CBF. As a result of the setbacks in the early years of the twenty-first century, particularly at the federal level, the foundation created a new federal affairs program "to make the Bay a national issue in the same way that the Everglades or the Great Lakes are national issues." CBF's pick to run the office was Doug Siglin, who had been with CBF since 2001 but had held a similar federal policy position with the Nature Conservancy and the World Wildlife Fund before coming to CBF. Although the strategic allocation of resources toward a DC office had been in the works for some time, the announcement of Siglin's new role happened to coincide with another round of cuts to environmental protection from the Bush White House. "An overall reduction of $33 million by the President in money directed specifically at bay restoration" underscored the challenges that Bay restoration faced at the federal level. Even CBF's education division, normally concerned with local and state-level policies, banded together with educators across the country in a coalition cheekily called No Child Left Inside to push for inclusion of environmental education in the reauthorization of Bush's signature No Child Left Behind legislation. Baker, using some of the strongest language in his three decades at the helm of CBF, swore "to force compliance with the historic federal/state Chesapeake 2000 Agreement" and promised to "turn the heat up even further" on what he called a "culture of failure," especially in the federal bureaucracy. The foundation normally framed litigation as an option of last resort, but by 2007 it was boasting that under Mueller's watch "staff attorneys used litigation more ambitiously than at any time in CBF history"; this included winning a major lawsuit that confirmed the standing of environmental groups to bring lawsuits in Virginia.[40]

CBF's tactics changed not only in the context of obstruction from the Bush administration and apathy at the state level but also in the midst of increased criticism of the foundation's role within the Chesapeake environmental movement. The political scientist Howard Ernst, writing in 2003, acknowledged that CBF was "the premier advocacy group for the Chesapeake Bay," but he proceeded to call "for the organization's leadership to rethink their political strategy for the Bay." Ernst's analysis of CBF is worth exploring in detail because it addresses most of the major critiques of the foundation's work. Ernst identified CBF as "a classic example of a group's organizational interests pulling it in one direction, toward political moderation, and its advocacy interests pulling it in another direction, toward political activism." CBF had long been criticized for being too moderate. For instance, in 1984 Greenpeace's national campaign director, Steve McAllister, had said that "they still do not push hard enough and their programs will *not* end up saving the bay. They have to have more aggressive tactics." While acknowledging that CBF's 501(c)3 status limited its legal ability to engage in political activities, Ernst pointed to the Sierra Club, which maintained both a political action committee and a separate legal defense fund, as a counterexample. In Ernst's view, CBF's decision not to pursue those options "severely limited its political relevance." Ernst's analysis echoed a frustration on the part of many in the environmental community. As he put it, "The Chesapeake Bay Foundation remains the only advocacy group with sufficient resources to launch a sustained political campaign for the Bay.... The future of the Bay depends on political action, and the Chesapeake Bay Foundation, as the region's leading environmental group, has a responsibility to be at the forefront of the political battle." While foundation leaders bristled at Ernst's analysis at the time, he correctly anticipated the direction the organization would take by the mid-2000s.[41]

Another important factor was that CBF was no longer one of just a handful of organizations advocating for the Bay. It had long been the dominant regional force in environmental issues, but the landscape had changed dramatically by the early years of the twenty-first century. For most of the Chesapeake Bay environmental movement's history, the other main group had been the Citizens Program for the Chesapeake Bay, later renamed the Alliance for the Chesapeake Bay. The Alliance was an even less confrontational group than CBF. According to its longtime director, Frances Flanigan, it "was not an advocacy organization, it was a consensus building organization." By the first decade of the twenty-first century the picture was changing rapidly; the growth of the Waterkeepers Chesapeake is a case in point. A member of the international Waterkeeper Alliance, the Waterkeepers Chesapeake emerged in 2004 with the help of the Chesapeake Bay Commission as a coalition of nine separate watershed groups. The oldest, the Chester Riverkeeper, started in 1986 and was the only organization of this type until the Potomac Riverkeeper began in 2000, followed shortly by the James Riverkeeper in 2001, with the Severn and Assateague both getting organizations in 2002, followed by several others. Another half dozen or so organizations would join shortly after 2004, with the total reaching sixteen in 2024. The growth in Riverkeeper organizations reflected an explosion of nonprofits across the watershed linked to some environmental cause or place; the Chesapeake Bay Program maintains an online list of "more than 600 organizations actively working to restore and conserve natural resources and create sustainable communities across the Chesapeake Bay watershed." These organizations range from groups as small and locally focused as the Rivanna Conservation Alliance—established in 2016 to clean and protect the Rivanna River, which runs through Charlottesville, Virginia, as part of its 42.1 mile course—to massive international organizations like the Nature Conservancy, which protects nearly a quarter million acres of priority lands in the Bay's watershed.[42]

The increase in the number of groups inevitably led to competition for funds and duplication of effort, but in many cases the proliferation of organizations across different levels created opportunities for greater impact. For instance, a project that planted native vegetation on more than a mile of streambank on a large cattle farm in Western Maryland involved more than a dozen organizations and one hundred volunteers, including funding from the National Fish and Wildlife Foundation, technical advice from the Frederick County Soil Conservation District, and logistic and volunteer support from CBF and the Monocacy and Catoctin Watershed Alliance. Indeed, the increase in the number of organizations reflected both a renewal of interest in the Chesapeake Bay and increased opportunities because of the sheer amount of work required to protect the Bay. Similar to how the watershed contained many ecological niches, the Chesapeake's nonprofit ecosystem grew to accommodate many organizational niches. Without dismissing the tension between newer organizations and CBF for funding, membership, and leadership over the direction of the Bay's restoration, this competition was made possible because the amount of public interest and financial support had grown significantly since the depths of the 1990s. Despite, or perhaps because of, the emergence of dozens of new watershed-focused groups and involvement from national environmental organizations, CBF's membership doubled from 100,000 in 2001 to 200,000 in 2008, while its revenue grew by 25 percent, from roughly $17 million to $21 million. Simply put, the Chesapeake environmental movement and its leading organization were bigger and better financed than ever before. By the middle of the first decade of the twenty-first century this larger, stronger, and broader environmental movement had begun to score some major policy victories. Unsurprisingly, one of the first major breakthroughs came in Maryland.[43]

BREAKING PARTISAN GRIDLOCK FOR THE BAY

Even after developing a more robust litigation program and embracing a more assertive strategy overall, CBF was not about to be confused with more radical, direct-action groups like Greenpeace or Earth First! Belief in working through the political system to create change was hardwired into the organization's DNA. Recall that some of its key founders included local and national politicians from both major political parties, and the group's first director, Jess Malcolm, earned his pink slip in part because he was perceived as too radical. Therefore, it should not come as a surprise that despite deepening toxic partisanship and tremendous frustration with the Bush administration, William Baker could still write in 2005 that "CBF has consistently believed that elected officials on either side of the aisle can do good, or bad, for the Bay. Our conclusion that the 'politics of postponement' have stifled progress for the last several years applies to both Republicans and Democrats. And as we have methodically worked to rebuild a stronger sense of political will for the Bay, we have been equally party-blind." This belief bore fruit at the state level, but the successes of CBF's approach in Richmond, Harrisburg, and Annapolis both underscored its failures to reach the Bush administration in Washington, DC, and highlighted the urgent necessity of federal leadership to preserve and restore the Chesapeake.[44]

One major reason Baker could still be so optimistic was that in May 2004 CBF and its partners had made a colossal breakthrough with Maryland's Republican governor, Robert Ehrlich, and the Democratic-controlled General Assembly to create a nearly $1 billion Bay Restoration Fund to pay for upgrades to sewage and wastewater infrastructure. The legislation was a triumph of the Bay's political ecosystem. The Chesapeake Bay Commission analyzed the most cost-effective strategies to reduce nutrient pollution and found that upgrading water treatment plants would be "the most reliable and immediate way" to cut pollution; by their estimates fully implementing the upgrades would achieve a third of the nutrient reduction goal for the Bay. With this goal, CBF took on the advocacy challenge, writing editorials, activating its grassroots network, commissioning a poll that showed support for the measure, and developing an interactive, online "sewage report" that allowed people to review the shortcomings of their local wastewater treatment plant. Most of all, CBF helped organize a broad coalition that included "the leaders of 13 prominent Bay fishing groups" and "activists, boaters, fishermen, watermen, Riverkeepers, environmental leaders, and citizens" to demand action on wastewater upgrades. Part of the measure's appeal was that it called for a flat, $2.50 monthly tax on every household in Maryland to ensure that every resident in the state did their part to help the Bay. But more importantly, support for the bill did not come from just one organization; it came from working people like charter boat captains, commercial fishermen, even seafood restaurateurs, alongside a range of new grassroots environmental groups demanding action on the Bay. Such broad support explains how a Republican governor who ran on an antitax platform ended up working with the Democrats in the General Assembly to pass a tax increase that took the biggest bite out of nutrient pollution since the phosphate ban twenty years earlier.[45]

As had so often been the case in the Chesapeake's environmental history, once Maryland took action, environmentalists were able to leverage increasing pressure on Virginia and Pennsylvania to do something too. In Pennsylvania the issue was put directly to voters in a May referendum shortly after Maryland's own legislation passed. Voters approved a $250 million bond for wastewater infrastructure, which was a mixed bag: only $50 million was earmarked for upgrading existing infrastructure; the remainder had to be spent on "water projects associated with economic development," which could include harmful new sprawl development. Nevertheless, Pennsylvania was on board, which meant that Virginia was the lone holdout. Ahead of the 2005 legislative session, CBF used the same "new strategies" that had worked in Maryland to launch "an aggressive campaign that included television, newspaper, and radio ads, public meetings, paid lobbying and strategic electronic communications" to help persuade the Virginia Assembly to make a $50 million appropriation to fund upgrades to sewage treatment plants. Unlike in Maryland, this was a one-time payment, not a new permanent funding mechanism. The commonwealth's legislators declined to follow Maryland's approach, but although they adopted different tactics, the end result was the same. In 2006 Virginia once again made a one-time payment, quadrupling its prior commitment with "a record $200 million . . . to upgrade nutrient removal technology at wastewater treatment plants." In 2007, benefitting from shifting political winds on environmental issues, Governor Tim Kaine proposed another $250 million through municipal bonds, cleverly called "Bay Bonds," which ultimately passed the Virginia Assembly unanimously. According to CBF, with more than $500 million appropriated from 2005 to 2007, "the action 'finishes the job' on addressing pollution from Virginia sewage plants." It had to be a bittersweet moment for many Chesapeake advocates. Yes, Maryland, Virginia, and Pennsylvania had finally agreed to spend more than $1.75 billion between them to upgrade wastewater treatment plants, but some must have let their thoughts drift back to the summer of 2001, when Senator Sarbanes pitched a plan for the federal government to pay for those same upgrades at less than half the cost, and wondered what might have been if the legislation had passed instead of stalling and then fading into obscurity. How much further along the road to recovery would the Bay have been? What other important goals could $1.75 billion in state dollars have funded? Those thoughts must surely have lurked at the margins even as environmentalists rightly celebrated their breakthroughs.[46]

Wastewater infrastructure was just one area in which Chesapeake environmentalists scored victories as public opinion soured on the Bush administration's antiscience, antiregulatory approach to environmental policy. In 2006 Maryland's Governor Ehrlich signed "the most rigorous power plant emissions legislation ever passed by any state in the country," largely in response to the Bush EPA's refusal to regulate coal-fired power plants. Later that year, the state blocked a massive development plan slated to destroy nearly eleven hundred acres of vulnerable forest, farmland, and wetlands abutting Blackwater National Wildlife Refuge and eventually purchased more than seven hundred acres of the plot to protect in perpetuity, while allowing development to proceed on 10 percent of the original land. In 2007 the state created another dedicated funding mechanism, the Chesapeake Bay 2010 Trust Fund, during a rare special session of the General Assembly. The fund provided another dedicated $50 million annually derived from existing gasoline and car rental taxes. Virginia, which had a history of reluctance to restrict commercial harvesting, agreed to keep sanctuary oyster beds off limits to harvesting by dredging and to cap the annual harvest of menhaden, an important species of filter feeder fish crucial to the Bay's food webs. Virginia also made it easier for local governments to enact land conservation programs. Other signs pointed to a new surge of enthusiasm for the environment; for instance, in 2007 CBF's annual Clean the Bay Day in Virginia drew more than six thousand volunteers, the largest turnout ever for the event.[47]

Pennsylvania did not do as much overall as Maryland or Virginia, but it did make one signature contribution to the Chesapeake antipollution effort. For once, on an important issue, "in the lead was Pennsylvania." Fittingly, that issue was agricultural pollution, and Pennsylvania's legislators deserved praise for coming up with an inventive tax-credit system known as the Resource Enhancement & Protection (REAP) program. If Pennsylvania's prior Nutrient Management Act was the stick, REAP was the $10 million carrot to encourage farmers to reduce nutrient and sediment pollution. REAP made tax credits available to farmers who made upgrades to their operations, such as fencing cattle out of streams or building storage pits for excess manure, for up to 75 percent of the cost of the project. The program had two important features. First, in order to be eligible for the tax credits, farmers needed to be in compliance with the Nutrient Management Act. Second, previous programs had offered financial support to farmers, but REAP's creative twist was that it made those tax credits transferrable from farmers to businesses. For example, a farmer who wanted to build a livestock crossing over a stream (thereby preventing shoreline erosion and keeping excrement out of the water) but lacked the capital to do so could find a business "sponsor" who would pay for the upgrade and receive the tax credit. REAP proved to be a popular program; as of 2020 it had supported improvements on more than three thousand farming operations, 75 percent of which were in the Chesapeake watershed. The only problem with the program was that it could not meet demand; according to the Pennsylvania Department of Agriculture, in every year of REAP's existence "the number of applications for available credits have exceeded the allocation of credits available for that year."[48]

REAP is an excellent example of the progress and the limits of Chesapeake environmentalism at the state level. In contrast to the years-long battle over the Nutrient Management Act in the 1990s, REAP passed in 2007, just one year after being introduced to the General Assembly. The bill benefitted from an unusually broad coalition of supporters. In addition to longtime green groups like the Chesapeake Bay Foundation and the Pennsylvania Environmental Council, REAP's supporters included newer grassroots groups like the Lower Susquehanna Riverkeeper, local businesses like central Pennsylvania's Nature's Best Organic Feeds, and perhaps surprisingly, industry associations like the Pennsylvania Farm Bureau and the Lancaster Chamber of Commerce. In all, nearly seventy organizations endorsed the measure, which helps explain its passage; however, this support was not enough to ensure funding at a level to meet Pennsylvania's obligations under Chesapeake 2000. Worse, there was no dedicated funding mechanism for REAP, which meant that funding for the program was slashed during the Great Recession. While REAP was a step in the right direction and remains a critical component of the Bay's recovery, it was no silver bullet. It took years for the cumulative annual improvements to add up to a significant reduction in nutrient pollution, which is still far short of the state's commitments. Had Pennsylvania begun REAP in 2001 or even 2003 or 2004 and given the program increased funding (which Pennsylvania finally did in 2019) and paired it with other measures, perhaps it would have achieved its Chesapeake 2000 goals. Undoubtedly, the Bay would have been in better condition sooner. Instead, the commonwealth, like its partners in the Chesapeake 2000 agreement, failed to make good on its promises. At least the states made a belated effort to scramble in the right direction, which contrasted sharply with the response from the federal government. In 2004, after Maryland broke the logjam on Chesapeake legislation, the Chesapeake Bay Commission sent a letter to George W. Bush asking him to build upon the restoration effort's successes by issuing an executive order "that identifies the Chesapeake as a national treasure and directs federal agencies to better coordinate, target and fund the actions needed to fulfill the federal commitments to Bay restoration." Bush never responded to the letter, but he did not need to, because his administration had already given a clear answer.[49]

FIGHTING FOR FEDERAL LEADERSHIP

The contrast between federal inaction on the Bay and the increasingly vigorous state-level responses only grew during George W. Bush's second term.

Everywhere one looked, except for the executive branch of the U.S. government, the Chesapeake Bay had supporters. In Maryland and Virginia, politicians of both parties needed to declare some degree of fealty to the cause to get elected; an array of nonprofits collaborated with citizens and members of the regulated communities, like commercial fishers and farmers, to come up with solutions to the Bay's poor water quality. Despite these developments, the scope of the Bay's problems required greater federal involvement; without it the restoration would remain hobbled. The Chesapeake Bay Foundation's longtime assumption that if the states led the way by increasing their spending on the Bay, the federal government would follow was being sorely tested. By Bush's second term, it was clear that no amount of leading by example, no paeans to bipartisanship, no amount of cajoling would lead his administration to live up to the pledges the EPA had made under Chesapeake 2000 or to enforce the Clean Water Act. There would have to be a fight.

CBF had been bracing itself for a fight at the federal level in the middle of the first decade of the twenty-first century by hiring Jon Mueller to run the new Chesapeake Bay Watershed Litigation Program and Doug Siglin to direct the foundation's federal lobbying effort, but once again events beyond the watershed shifted the fortunes of the Bay restoration movement. This time, the historic midterm sweep by Democrats in 2006 gave the Chesapeake Bay a boost. The first major victory for the Bay came in 2007, when Congress overrode a Bush veto to enact the $23 billion Water Resources Development Act. It was the first time a Bush veto had been overridden, and there was significant Republican support for the override. While one GOP holdout, Senator Jim DeMint of South Carolina, lamented that "sadly, because the authors of this bill have rained a few earmarks to every member's district, Congress didn't have the courage to stop this reckless overspending," Chesapeake environmentalists rejoiced in soaking up hundreds of millions of dollars from the blue-wave backlash to Bush.[50]

The Water Resources Development Act was a relatively easy win, but the bigger fight CBF and other groups were gearing up for was more than twenty years in the making: a historic reallocation of funds in the federal Farm Bill to the Chesapeake region. The massive, quinquennial omnibus agricultural spending bill had long been an elusive CBF target. In 1985, still riding the high of the first Chesapeake Bay Agreement, CBF staff testified before the House Agricultural Committee. CBF's vice president at the time explained the organization's interest in the matter because of the emerging understanding of agricultural pollution's impact on the Bay and because "there is also an obvious policy role to be filled by CBF, since few farm organizations address environmental issues." Staff testified for provisions that were environmentally friendly, but there was no strategic effort to use the Farm Bill to fund pollution reduction activities. However, in 2001 William Baker declared that "making the Farm Bill more supportive of agricultural conservation efforts is a top CBF priority." CBF worked with Representative Wayne Gilchrest (R-MD) to craft an amendment that "would have shifted billions of dollars [nationally] from traditional farm subsidy programs . . . to grants to help small farmers reduce polluted runoff from fields, provide habitat for wildlife, and resist urban sprawl." The amendment had wide bipartisan support, including cosponsors Ron Kind (D-WI), Sherwood Boehlert (R-NY), and John Dingell (D-MI), but it was narrowly defeated in the House, 222–200. What made this defeat sting for CBF was that the nay voters included thirteen representatives from Maryland, Virginia, and Pennsylvania. Had CBF been able to construct a voting bloc for the Bay's farmers, the measure would have passed.[51]

In 2006, with reauthorization of the Farm Bill looming, CBF's Doug Siglin led a switch in tactics to lobby aggressively for Chesapeake-specific funding instead of supporting nationwide conservation policies. CBF was no longer the lone organization advocating for environmental issues in the Farm Bill; it had spent years building alliances with farm communities (like the one that helped pass REAP), and it was clear to both environmentalists and the region's political leadership that the Farm Bill "represents this region's best opportunity to substantially *scale* up agricultural conservation activities." The governors of Maryland, Virginia, and Pennsylvania issued a joint resolution calling on Congress to do more through the Farm Bill for the Bay, and congressional allies Chris Van Hollen (D-MD) and Bobby Scott (D-VA) in the House and Barbara Mikulski (D-MD) in the Senate answered that call by introducing a marker bill, the Chesapeake Healthy and Environmentally Sound Stewardship of Energy and Agriculture Act (CHESSEA), the acronym a nod to the Bay's mythical counterpart to the Loch Ness monster, Chessie.[52]

The grueling two-year fight to get CHESSEA's provisions included in the Farm Bill "was the most intense ever waged at the federal level by the Chesapeake Bay Foundation." It was also highly controversial as many environmental groups objected to the bill because of its overall support for the fossil fuel–driven industrial agricultural monocultures, which receive the bulk of the Farm Bill's billions in federal spending. Nevertheless, Siglin and CBF staff framed a creative campaign that put farmers at the center of the debate, sometimes to the consternation of other environmentalists. Stressing regional inequity, the foundation called for the Farm Bill to "give Chesapeake-area farmers their fair share of federal dollars," emphasizing that Chesapeake farmers (including those in Pennsylvania) received proportionally far less federal funding than the average American farmer. At the time, farmers in the watershed only received federal support worth about four cents per dollar of farm production compared with the national average of nine cents per dollar. The greatest beneficiaries of federal largess were farmers who grew cotton, rice, corn, wheat, and soybeans, which were not major crops in the watershed. It was an uphill battle, but thanks to policy coordination from the Chesapeake Bay Commission, broad public support for the Bay, and large Democratic majorities in Congress, the Farm Bill finally passed in 2008—once again over a Bush veto—with $440 million earmarked for Chesapeake farmers over the bill's five-year lifespan. It was "the largest-ever federal investment in on-the-ground pollution reduction for the Bay states." It also turned out to be an effective investment. At the time of the bill's passage, the expectation was that if fully implemented, the provisions could reduce nitrogen pollution by 40 million pounds annually, or roughly one-third of the Chesapeake 2000 goal. In 2013 the U.S. Department of Agriculture published a report on the impact of conservation practices in the watershed during the lifespan of the Farm Bill and found that conservation practices had reduced nitrogen pollution by 48 million pounds, among a host of other economic and environmental benefits. The report concluded that "historic levels of conservation implementation are achieving unprecedented results in the Chesapeake Bay region." As with REAP, this example is bittersweet because it highlights how much pollution reduction the region could have achieved had there been substantial federal support earlier.[53]

By 2008 federal leadership had finally arrived, at least with respect to funding from the legislative branch. Both major federal spending bills, the Water Resources Development Act and the Farm Bill, were boons to the Chesapeake's recovery, but they depended on Democratic majorities strong enough to override the veto of an unpopular president with the help of some bipartisan support. If it was not already clear that federal leadership and federal funds might well be fleeting, looming behind these two major victories was the Great Recession of 2008, which threatened to eclipse Chesapeake and national environmental policy just as the war on terror had done earlier in the decade. Despite the stakes, CBF maintained its strict nonpartisan approach in the run-up to the 2008 federal elections. The organization ran a mock campaign featuring "Captain John Smith" for president, with a live actor playing Smith. In announcing his "campaign," the organization released a statement that "CBF does not endorse candidates. CBF is running a fictional candidate, Captain John Smith, to elevate the Bay and clean water in the presidential election." This was undoubtedly disappointing to many who wanted the region's largest environmental group to take a firmer stance on the issue, but CBF was playing a longer game.[54]

While "Captain Smith" was out stumping for the Bay, Jon Mueller and CBF's legal team were unleashing a plan they had been crafting for the past three years. On October 29, 2008, CBF notified the EPA of its intention to sue the agency for failing to enforce provisions of the Clean Water Act that would require a "pollution diet" for the Chesapeake Bay. CBF's partners in the case indicated the breadth of the Chesapeake environmental coalition, even if some of those partners might have rejected the label "environmentalists." The major watermen's associations in both Virginia and Maryland joined CBF, alongside their sometimes rival, the Maryland Saltwater Sportfishermen's Association, and key political leaders, including former Maryland governor Harry Hughes, who had signed the original Chesapeake Bay Agreement, Maryland state senator Bernie Fowler, former DC mayor (and former CBF trustee) Anthony Williams, and former Virginia natural resources secretary Tayloe Murphy (also a former CBF trustee). Larry Simns, the longtime president of the Maryland Watermen's Association, said that environmentalists, political leaders, and commercial and recreational fishers had joined forces "because we're backed into a corner. We've all been preaching to clean the Bay up, with no results. . . . We're at a crucial point here, and unless we do something now we're going to lose the Bay completely."[55]

What CBF, Simns, and others wanted was for the EPA to develop and enforce a Total Maximum Daily Load for nitrogen, phosphorus, and sediment entering the Chesapeake Bay under Section 303(d) of the Clean Water Act. At its core, a TMDL is just what the name implies, the maximum amount of a pollutant that can enter a body of water without degrading it to the point where it would appear on the EPA's impaired-waters list. This was not the first time CBF had called for a TMDL; in 1997 the foundation asked the governors of the Bay states to "implement the portion of the Clean Water Act which requires that for all waters designated as 'impaired' a total maximum daily load (TMDL) is developed for the pollutant of concern." Interestingly, CBF made this call in the context of reducing toxic pollution in the Bay, not nutrient pollution. Later that summer, CBF notified the EPA and the state of Maryland that it would sue them to "seek the required monitoring and maximum daily loads to help clean up Maryland's waters, including the Bay." Before that case went anywhere, *Pfiesteria* struck the region and initiated the chain reaction that led to the regional adoption of most of CBF's goals in Chesapeake 2000. Notably, the signatories to the agreement acknowledged that "we have agreed to the goal of improving water quality in the Bay and its tributaries so that these waters may be removed from the impaired waters list prior to the time when regulatory mechanisms under Section 303(d) of the Clean Water Act would be applied." Chesapeake 2000's adoption stayed CBF's lawsuit, but when it became apparent in the early years of the twenty-first century that implementation of the agreement was a hopeless case, CBF hired Jon Mueller and began plotting how to compel enforcement of the Clean Water Act.[56]

What Jon Mueller was contemplating in the middle of the first decade of the twenty-first century was an ambitious and creative interpretation of both the Clean Water Act and multiple rounds of Chesapeake Bay Agreements. Under the Clean Water Act, TMDLs only apply to point source pollution, but without a mechanism to compel reduction in nonpoint source pollution, most importantly from agriculture, the Bay would not recover. Mueller saw an opportunity to use the legacy of Chesapeake Bay Agreements to address this problem. In an interview some years later, Mueller recalled that suing the EPA

> allowed us to bring kind of a novel legal theory, not just under the Clean Water Act, but under a federal, kind of common law. . . . I looked at the Chesapeake Bay Agreements and I said, "These are signed by government officials all saying 'we are gonna do X' and the public reads those things, and so great, they're gonna do X and it's really gonna fix the Bay, which is

exactly what I want them to do." Ten years go by, they don't do it. They write another agreement. Same thing. Don't do it. Don't do it, don't do it. But they keep signing these agreements.... Why is it that we can't turn to our elected officials and say, "You signed this document and you were going to do X by this date and you didn't do it." So, I did a lot of research to get into this interstate compact theory in which states basically contracted with the federal government to clean up the Bay and they weren't doing it, so we alleged that theory in our complaint.

Mueller added, "It took us three years to develop that litigation. And that was with a lot of thought here throughout the organization on a lot of different levels." By 2008 the attitude toward the EPA was that "we're kind of at our wit's ends with you guys, nothing's happening, there is no Bay TMDL, there is no real plan for meeting the Clean Water Act objective," and regardless of who won the presidential election, foundation leadership felt it necessary to threaten the agency with legal action. Mueller admitted that "we could have taken a wild guess at what the Obama Administration was going to do and not sued in January," but CBF and its allies felt that they could not wait any longer for the EPA to do something for the Chesapeake. Thus, despite statements of support by Obama and his proxies, CBF and its allies, with Bernie Fowler as the lead plaintiff, sued what was technically still the outgoing Bush administration's EPA on January 5, 2009.[57]

The coplantiffs' sense of urgency was well founded. Although the Obama administration was supportive of the Chesapeake Bay Program, it still took a year and a half of negotiations to reach a consent decree in May 2010. The EPA agreed to establish a Bay-wide TMDL by December 31, 2010, that would account for both point and nonpoint source pollution. Unlike past Bay agreements, this consent decree forced the EPA to accept leadership responsibility for the pollution reduction goals. When the EPA finally released the Chesapeake Bay TMDL just two days ahead of the deadline, it was, in William Baker's words, "the holy grail of Bay restoration." It was also immediately challenged in court by the American Farm Bureau Federation and the Pennsylvania Farm Bureau. The onetime allies on the federal Farm Bill had become opponents of the Bay restoration's most important legal tool. They were joined by a host of agricultural industry groups and the National Association of Home Builders. Detailing the course of legal action is well beyond the scope of this study, but a few brief remarks will suffice to close the case on the Bay TMDL. Thanks in no small part to the legacy of agreements between the states and the EPA, Judge Sylvia Rambo, of the District Court for the Middle District of Pennsylvania, upheld the legality of the Bay TMDL in 2013, and the measure survived appeal to the Third Circuit Court in 2015. Notably, the appeal was joined by amicus curiae briefs from the attorneys general of twenty-one states. The Bay was undoubtedly in the national spotlight as the case went in 2016 all the way to the United States Supreme Court, which in a 4–4 ruling denied the case a hearing, allowing the Bay TMDL to stand. It was in many ways a decision more than thirty years in the making.[58]

At the same time that CBF and its allies were suing the EPA, they were also negotiating with the Obama White House for another long-sought component of federal leadership, an executive order formally recognizing the Chesapeake as a national treasure and ordering the EPA to coordinate the many aspects of the federal government's role in the Chesapeake's recovery. The idea was first broached by the Chesapeake Bay Commission in 2004, but it went nowhere under the Bush administration. Despite CBF's longstanding position that any politician from any party could be influenced to do right by the Bay, "with the election of the new President . . . the Chesapeake Bay Foundation knew it had a once-in-a-generation opening to change the course of Bay history." The Obama administration was more receptive to environmental protection than its immediate predecessor, and the contrast heightened the sense that Obama's presidency was an opportunity that could not be missed. Thus, even before Obama's inauguration, there was an "all-out offensive . . . to convince President Barack Obama and the new Congress to make jump-starting the long-delayed Bay cleanup a top federal priority." Tim Kaine, then the governor of Virginia and chair of the Chesapeake Bay Program's executive council, played a crucial role in keeping the issue in front of the new EPA administrator, Lisa Jackson, as did CBF's lawsuit, delegations from the Chesapeake Bay Commission, and other political leaders, along with other local and national environmental groups. Chesapeake environmentalists got a much swifter response than in the case of the TMDL. As noted at the beginning of this chapter, on May 12, 2009, Barack Obama signed Executive Order 13508, which opened by declaring that "the Chesapeake Bay is a national treasure" and that "restoration of the health of the Chesapeake

Bay will require a renewed commitment to controlling pollution from all sources.... The Federal Government should lead this effort."[59]

Obama's executive order was important both symbolically and substantively. Symbolically, calling the Bay a national treasure appealed to environmentalists who wanted the estuary's status elevated to a level be on par with that of some of America's greatest national parks. Although Executive Order 13508 did not quite do that, it was an important step in the right direction. The order also led to substantial changes for federal oversight of the Bay's restoration. The EPA was the primary federal agency responsible for the Bay, but numerous other agencies made decisions that affected the Bay (such as the Department of Agriculture through the Farm Bill), and technically only the EPA was a party to the Chesapeake Bay Agreements. As was often the case in such situations, agencies with overlapping responsibilities did not coordinate and cooperate with one another unless explicitly ordered to do so. Thus, E.O. 13508 created yet another body, a Federal Leadership Committee, chaired by the administrator of the EPA and including senior representatives from the Departments of Agriculture, Commerce, Defense, Homeland Security, the Interior, and Transportation. In addition to giving the EPA the authority to oversee and coordinate the implementation of programs affecting the Bay, Executive Order 13508 also instructed the committee to produce a list of specific goals and actions that would restore the Bay. In many ways, this was reinventing the wheel. Chesapeake 2000 contained a perfectly adequate set of goals and actions that, if implemented, would have made significant improvements to the Bay's water quality and living resources. Executive Order 13508 was an improvement because it gave the EPA a clear leadership role, but the fact that it was building upon previous agreements and that any subsequent president could undo the action with a stroke of a pen combined to fuel CBF's continued insistence on a legally binding settlement for the Chesapeake Bay TMDL.

CONCLUSION

The Chesapeake Bay TMDL and Executive Order 13508 joined a host of state and federal measures that made the period 2008–10 an inflection point in the Chesapeake Bay's environmental history. Compared with the darkest moments of the 1990s, these years marked a profound and fundamental shift in the nature of the Chesapeake Bay restoration movement. Because the struggle to restore the Chesapeake has not ended, there is a temptation to carry this historical narrative closer and closer to the present day. That temptation has already been well enough indulged by following the TMDL case to its conclusion. Obama's executive order is a fitting place to conclude this narrative because it marks an evolution in the Chesapeake Bay Program and closes a distinct era in Chesapeake history. One of the reasons why Arthur Sherwood, Porter Hopkins, and their friends from the Annapolis Yacht Club started the Chesapeake Bay Foundation was that Congressman Rogers C. B. Morton told them that significant assistance from the federal government would not be forthcoming. As Sherwood recalled CBF's origin story a decade later, Morton "caught the ear of his neighbors with the disarmingly frank admission that neither he nor the government at large for which he worked should be expected to 'solve all the problems of the Chesapeake'; that it was high time for the 'average' citizen to do something for himself. Thus was CBF born. One spark among many ignited. A chance encounter between a public official and his constituents brought a new organization into being." More than forty years later, CBF played a definitive role in forcing the U.S. government to at long last assume a leading role in the protection and restoration of the nation's estuary. Whether federal leadership continues to be a key component of the Chesapeake's recovery and whether those efforts will be successful are not yet questions for a historian to answer, though they will be one day.[60]

Part of what makes writing the Chesapeake's history so challenging is that the long-term fates of many of the key components of the recovery are not assured. Environmental factors from climate change to invasive species could undo decades of hard and expensive work. As noted, executive orders can easily be undone, the conservative legal movement is actively seeking to weaken federal environmental regulations, and support for public spending on the Bay is always one election away from catastrophe. Even the Chesapeake TMDL has its vulnerabilities. CBF's Mueller calls enforcement "probably the weakest link in the chain" because "there's no mandatory obligation that the TMDL ever be implemented." The federal government can withhold funding for failure to implement a TMDL, but states could refuse federal dollars and resist TMDL implementation. As Mueller observed with some prescience in 2012, "We're going to have to be really creative in how we go about making certain that the TMDL is in fact implemented."[61]

The history laid out in these pages certainly suggests that there will be challenges to implementing the measures necessary to continue the Bay's restoration and remove it from the impaired-waters list. Nevertheless, this history also reveals that in the face of growing population pressure and at times apathetic political leadership, a popular movement grew that has helped to sustain the Chesapeake as a place capable of evoking all the wonder and magic of nature while still supporting age-old ways of living off the land and water. This is a tremendous accomplishment. By 2010, as it transitioned into a new era, the regional environmental movement had reached its strongest position yet. Measured by the growth in finances and membership at CBF, or the growth in the total number of organizations involved with the Bay inside the watershed, popular support for the Bay had never been greater. Measured by the growth in federal and state spending on the Bay, more money had never been available for restoration. Perhaps most of all, despite the challenges of measuring the ecosystem's health, by 2010 the Bay showed tentative signs of progress. Although the Bay had not returned to where it was before Tropical Storm Agnes devastated the region, it was healthier, and the environmental movement's ability to meet threats to the Bay was greater than ever. The Chesapeake will face threats in its future, and it will likely endure setbacks as well as successes, but if the history in these pages is any guide, it will always have energized, creative, and dedicated people fighting to save the Bay.

CONCLUSION
LOOKING BACK, LOOKING AHEAD

The main narrative of this book concerns events that took place roughly between the mid-1960s and the mid-2010s. In short, the argument presented in these pages is that during this half century the Chesapeake Bay evolved from being an environmental issue fought over by individual states, namely, Maryland and Virginia, to a regional project that incorporated Pennsylvania with some coordination from the federal government, to a truly national environmental policy issue. While individual policies at the state and local levels of government remain crucial to the Bay's health, only the federal government has the resources and regulatory power to lead the Bay's recovery across the length and breadth of its six-state watershed. Although unfinished, the work to protect and restore the Bay chronicled in these pages is a clear success story in the annals of environmental history. Reaching this conclusion requires reconciling two seemingly contradictory positions. On the one hand, a set of policies enacted across multiple levels of government that restored and improved components of the estuary despite continued population growth and resource usage in the region is a triumph. On the other hand, the signatories to the Chesapeake Bay Agreements have failed to fulfill their promises, and the environmentalists advocating for the Bay have not achieved their stated goal of a broader, systemwide recovery. To evaluate the legacy of the Chesapeake Bay effort and consider lessons from this history for the future requires grappling with the truth of both statements.

Many people involved with the restoration effort would look at the Chesapeake Bay's current state and judge any conclusion of success to be a grave misinterpretation of the data. Gerald Winegrad, a former Maryland state legislator and a deeply respected, longtime leader of Chesapeake advocacy, is one such person. In a 2023 editorial about "the abject failure to meet the 2025 deadline for TMDL reductions," Winegrad eloquently described the disappointment many Chesapeake environmentalists felt at the state of the Bay's recovery forty years after the ink dried on the first Chesapeake Bay Agreement. Contrasting the expectations of 1983 with the reality of 2023, Winegrad wrote, "Those bright-eyed optimistic witnesses at the 1983 signing would now be hugely disappointed and alarmed, as am I. If environmentalists had drummed up a doomsday scenario for failing to take the necessary actions for Bay restoration, we now have arrived at that nightmare scenario." According to Winegrad, this "nightmare scenario" was a result of "repeated failures to rein in agricultural and development pollutants." Although Winegrad acknowledged some successes, his overall conclusion was that "the house of cards that was the Bay restoration plan has collapsed, and we are left with only broken promises and a Chesapeake facing a bleak future." Winegrad's analysis is a bit hyperbolic, but with the Bay restoration on track to fail to meet another recovery deadline in 2025, it cannot be dismissed when evaluating the movement's legacy.[1]

Christine Keiner wrestled with how to evaluate the mixed results of public policy in her excellent environmental history of the Maryland oyster industry. While she lamented that "environmental historians often have the depressing job of pointing out past mistakes," she emphasized that "it is equally crucial to recognize the value of what has worked." Following Keiner's lead, the argument here has not shied away from calling out past failures, but the weight of evidence does not support Winegrad's grim conclusion. The Bay restoration movement has done much of great value for the Bay and the many communities that depend on it. Recognizing what has worked—as well as what has not—will be essential to the future of the estuary. In his book on the Bay's restoration Tom Horton, the dean of Chesapeake environmental journalism, used a particularly eloquent metaphor of climbing a mountain to make a more nuanced point:

A parable for our times: Imagine you have vowed to climb a mountain. To make it, you need to reach a way station, not that far from the top, by a certain hour. The hour comes, and you are well short of your objective. Disappointing, but no disaster, you figure. You have covered a lot of ground, will make the way station soon enough, and will eventually reach the peak. Or so it seems. But now, with each step upward, the incline becomes steeper. And there has been a dismaying error—the way station is not nearly as close to the top as your map said. And the breeze is becoming a headwind.... Unfortunately, this parable is not about mountain climbing, nor is it one bit fanciful. It is about how all of us in the bay region, despite moving in the right direction, are failing our pledge to restore the health of the Chesapeake Bay.

The metaphor remains as relevant and compelling today as it was in 2003. It is impossible to deny the very real progress made to improve the Bay and equally impossible to ignore the glaring warning signs about the Bay's health. Describing the Bay's health on a year-to-year basis is a dizzying endeavor. Does the fact that Chesapeake oyster landings are reaching levels last seen in the 1980s indicate progress in recovery or a dangerous overharvesting of the fishery? Is the fact that the juvenile index for striped bass remains stubbornly low a sign of alarm, or is it just an indication that conditions have not been right recently for a dominant year class? Are pollution reduction efforts finally leading to a decrease in the size of the Chesapeake's dead zone, or has there just been a run of unusually dry summers? These and other questions are vital and urgent, but they cannot yet be fully answered.[2]

It is easy to lose sight of the big picture and get caught up in the almost daily torrent of news about the Bay's health. History can provide a more durable perspective that is not influenced by swings in yearly rainfall, harvest numbers, or "report cards," which may or may not be the most accurate measures of the ecosystem's long-term status. While the historical record certainly shows the movement's shortcomings, what emerges most clearly is a tradition of collaboration, coalition-building, and obstacles overcome to build a powerful political movement that has only recently come of age. The value of a deep historical analysis comes from being able to contextualize the successes and failures as products of enduring patterns—positive, encouraging factors like widespread support for the Bay and more negative, discouraging factors like the reluctance to enforce cleanup mechanisms. Moreover, a broader study of history reveals that structural obstacles are not unique to environmentalism. A litany of causes from civil rights and social justice to alleviating poverty and public health face similar degrees of incomplete progress marred by infuriating setbacks. One of the motivating factors behind the present volume is the hope that by understanding the historical origins of these problems, their evolution, and various attempts to address them, Bay advocates may find inspiration, hope, and endurance in the knowledge that though the journey may be long and milestones infrequent, the mountaintop is getting closer. With these questions in mind, let us review the path climbed so far.

FROM A STATE TO A NATIONAL ISSUE

One of the most compelling reasons to understand the history of the Chesapeake Bay restoration effort as a progression from a state to a national issue is that it illuminates the source of much of the frustration that environmental advocates like Winegrad feel. The persistence of the Bay's problems can make it seem as though all the work was for naught. Ann Swanson, the longtime director of the Chesapeake Bay Commission, told a revealing anecdote. When she was still working for CBF as a grassroots coordinator, she would go to schools and give presentations about the Bay cleanup effort. During the question-and-answer session at one of these assemblies, a student earnestly asked her what she was going to do for a job after the Bay was saved. To current eyes, this was a bit naïve, but as Swanson explained, "Nobody, nobody realized it would be so hard." Perhaps to a pessimist, the prospects for the Chesapeake's recovery might seem hopeless, but one benefit of historical perspective is greater appreciation for accomplishments in the face of tremendous obstacles. In our interview, Swanson was proud of the progress that had been made, but she left no doubt that the movement had not achieved its goals and that "if we accept, 'well, we made good progress,' then we're accepting a degraded Bay." Yet there is a huge distance between accepting some progress as an end state and appreciating it as a point on the way to an eventual goal.[3]

Another benefit of studying the past is knowing that one is not alone in facing these challenges. The frustrations of halting progress seem endemic to the work of environmentalism. In 1970 Jess Malcolm was devastated at the failure of the general public to support the Chesapeake. He wrote to CBF's board of directors that he was so "convinced that mankind is now on a self-destruct mode . . . that I languor in a mood of perpetual disbelief and frustration that others, particularly those who literally have the power of environmental life or death over us, are either unwilling or unable to recognize it." For Malcolm, his tenure as CBF's first director had been "fraught with profound disappointment." Although he does not receive nearly the credit he deserves for being one of the first and foremost Chesapeake environmentalists, it was perhaps for the best that he left the limelight in early 1971, because the challenges would only become greater. Toward the end of the decade, Malcolm's rival and successor as head of CBF, Arthur Sherwood, broke from his usual sunny optimism to "dwell on [CBF's] difficulties" and complain that "issues are crooked or, at best, meandering and curvacious, [sic] never straight to a point" and that more often than not a possible solution to a problem "slips imperceptably [sic] away and coalesces as just another insurmountable problem." These frustrations were not unique to environmentalists. As a working waterman, Larry Simns would have rejected being labeled an environmentalist, but he and the Maryland Watermen's Association, which he led for four decades, were involved to some degree with most major initiatives for the Bay. As Simns recalled in his semiretirement, he had been so frustrated at the challenges of running the organization, keeping watermen in line, and being their voice in Annapolis that "by the late seventies, I informed the board of directors of my plan to resign so I could get back on the water to fish full time." The MWA board refused to accept Simns's resignation and talked him into staying on. Simns would learn to "endure severe pushback and even rejection during the day from leaders while they were on stage," while privately continuing to argue for policies in a one-on-one setting. The experiences of Malcolm, Sherwood, and Simns are cited here not to trivialize current frustrations but to contextualize them as an enduring feature of environmental policymaking.[4]

As frustrating as the slow pace of progress was, giving up would have been much worse. At the time of Malcolm's complaints in early 1971, CBF had been formally incorporated for about four and a half years, and it had made little headway getting Maryland and Virginia to cooperate. It was not until Tropical Storm Agnes devastated the region that Charles Mathias was able to use the mounting concerns about the Bay's failure to recover to push through appropriations for the EPA study that would grow into the Chesapeake Bay Program. As argued in these pages, the evolution of the Chesapeake Bay Program from an EPA research study to a regional partnership in 1983 was not the start of the Bay restoration movement but rather a significant turning point. Although many participants from the 1980s recall the sense of optimism, and at times euphoria, surrounding the first Chesapeake Bay Agreement, others were less convinced. The editorial board of the *Daily Press* was skeptical, wondering if the Bay agreement, "for all its sound and fury, will have signified nothing." As they saw it, "for all the extravagant hoopla and dramatic proclamations, the single basic issue—is restoration of the bay a regional or a national problem?—remained unresolved." This was a remarkably astute observation, and one that bedeviled the Chesapeake restoration for another quarter century. The Bay states fought for the estuary, sometimes valiantly, sometimes halfheartedly, but rarely with the federal government as a full partner. Recall that the Chesapeake Bay Commission found that during the twenty years following the first Chesapeake Bay Agreement, only 18 percent of the restoration spending came from the federal government. This was akin to fighting the Bay's pollution woes with one hand tied behind one's back, or to use Horton's metaphor, trying to climb up a mountain in sandals instead of proper hiking boots.[5]

Federal leadership and funding is still not as robust as it needs to be, but the Bay is closer to receiving the support its stature as an ecological jewel of the nation should command. While the Bay Commission's 2004 bid for George W. Bush to issue an executive order to this effect failed, the commission saw success in 2009 with Barack Obama's Chesapeake Bay Protection and Restoration executive order. Funding for the Bay surged during Obama's presidency, remained surprisingly strong despite the vicissitudes of the Donald Trump administration and the COVID-19 pandemic, and continued at historically high levels during the administration of Joe Biden. Since 2014 the federal government has spent on average more than $500 million annually on the Bay. The total spending from the watershed states, nearly $1.2 billion, has been more than double federal spending, which indicates that there is still much room for growth in the federal share of the Chesapeake spending pie. Nevertheless, the Chesapeake Bay has never before experienced such a sustained run of federal support, spanning more than a decade and three presidential administrations. A 2013 analysis from the *Capital Gazette* (Annapolis) underscores this point. After exhaustive reporting, the *Capital* found that federal, state, and private partners had spent at least $15 billion to restore the Bay since 1983. In the decade since that report, those same partners have spent at least another $15 billion. Money is not the whole story, because it must be spent wisely for the Bay's recovery to continue. The point here is that the newest phase of the Bay's restoration is scarcely a decade old, and yet it has survived historically turbulent times in American politics and delivered nearly as much financial support to the estuary as was delivered during the previous thirty years. This achievement merits serious reflection and appreciation.[6]

Dividing the Bay's restoration into three phases—before the turning-point year of 1983, the thirty years between 1983 and 2013, and the time since—illustrates the evolution of the Chesapeake from a state-level concern, to a regional concern, to an emerging national priority. While acknowledging the work that remains to be done and the fact that the Bay remains impaired as restoration goals have not been met, the conclusions reached in these pages are as follows: The Bay languished as a territorial concern of Maryland and Virginia until the mid-1960s, when CBF and a few allies began to feebly push for a regional approach. It took an epic disaster and a decade of concerted lobbying by scientists, watermen, environmentalists, and their allies in state and federal office to reach a tentative, regional agreement, the first Chesapeake Bay Agreement in 1983. Although better coordinated, the state-level actions of Maryland, Virginia, and Pennsylvania were not enough to rapidly improve the Bay, and attention waned. With shifting national winds, the Bay restoration project was blown severely off course until 1997, when the mysterious terror of *Pfiesteria* reignited a new round of efforts at the state level. After setbacks in the early twenty-first century, the movement rebounded with victories in all three branches of the federal government that have led to subsequent improvements in the Bay's water quality and living resources. Whether future historians will choose to date a new phase of the Bay restoration from Obama's executive order in 2009, the EPA consent decree in 2010, the surge in funding since 2013, or the final legal challenges to the TMDL being vanquished at the U.S. Supreme Court in 2016, it is clear that a new era has indeed begun.

THE NEXT FIFTY YEARS

Most accounts of the Chesapeake restoration reference 1983 as the starting point, but as this work has argued, a deeper and richer history of the Chesapeake's restoration lies beyond 1983. Historians break the seamless flow of time into discrete periods to give it order and meaning. We will consider the period from CBF's incorporation in 1966 to the final defeat of the TMDL's opponents by the Roberts Court in 2016 the first half century of the modern Bay restoration movement. We are nearly a decade into the movement's next fifty years, and it is tempting to speculate on what the Bay could look like in 2066. This would be a fool's errand. History does not allow neat and tidy projections about what will happen in the future. Certainly some trends seem obvious: the EPA's authority to clean up the Bay is deeply threatened by the conservative turn of the U.S. Supreme Court, Maryland will lead in Bay policies most of the time, Pennsylvania will lag most of the time, climate change will pose unforeseen challenges, public support for the Bay will remain high. But the value of history is that it offers collective wisdom beyond the experience of any individual lifetime. Historical perspective can help us to think, feel, and act in response to the coming known and unknown challenges by learning from how others faced them in the past. History is ballast for stability against the crashing waves of time.

In the case of the Chesapeake, the newest phase of the restoration coincides with a new group of leaders in the watershed. At CBF, Hilary Harp Falk took the reins from William Baker in 2022, marking CBF's first leadership change since 1981. On the other side of Annapolis, in 2023 Anna Killius succeeded Ann Swanson, who had served as the CBC's director since 1988. This new blood has been a source of energy and encouragement, but as with all transitions, it makes for a precarious moment. In an interview after her first year on the job, CBF's Falk said that "we're all grappling right now [with] this big transition in the Chesapeake Bay movement, with new leaders, at a critical moment for the cleanup." Looking ahead to the next iteration of the Chesapeake Bay Agreement, Falk said, "One of our challenges is that we have really defined the Bay cleanup based on nitrogen, phosphorus and sediment. Now we have an opportunity to look more broadly at a number of other issues. As we are updating the Chesapeake Bay Watershed Agreement, that's a huge opportunity to look past nitrogen, phosphorus and sediment into other issues and really redefine what it means to save the Bay." Although some may groan at the prospect of yet another Bay agreement, the weight of historical analysis suggests that while flawed, the Chesapeake restoration effort has succeeded in forging a political consensus that the estuary is indeed a national treasure worthy of protection and that the federal government must lead the next chapter in the Bay's recovery. This has been no mean feat, and in the next fifty years this accomplishment must be remembered, celebrated, and defended.[7]

To continue the work of Bay restoration, it is vital to remember that first and foremost the prior fifty years of the restoration effort was about forging a political consensus. It is easy to pillory the various Bay agreements for failing to restore the Bay because they have indeed failed to restore the Bay. However, if we only think of these agreements as tools of a science-driven restoration, we will be perpetually disappointed. Knowledge is incomplete, many variables beyond our control can afflict and affect success, and to maintain and protect even a "saved" Bay in some glorious future would require constant vigilance. Likewise, if we think of the political leaders who make these agreements as doing so because it is the right thing to do, we will be perpetually disappointed as well. The Bay may only ever get a handful of political leaders in the mold of Charles Mathias. Instead, our politicians require constant poking and prodding to respond. This is deeply annoying, but it is not new nor unique to environmental issues. However, as with a Gestalt image, we can look at this history and reach a very different conclusion, one perhaps better suited to guide the Bay's recovery effort for the next fifty years. If we regard the agreements and the Chesapeake Bay Program as primarily tools for forging and maintaining a political consensus, the picture flips. Annoyance and irritation, instead of being in the foreground, fade to the back, and what emerges is a picture of perseverance and endurance. Against the full weight of the antienvironmental officials of Reagan's first term, the Chesapeake Bay Agreement emerged in 1983. Environmentalists successfully demanded that a new wave of leaders recommit to the Bay's recovery in 1987 and then again in 2000 and 2014. These accomplishments were not trivial, nor were they foregone conclusions. This repeated string of reaffirmations came from both Republican and Democratic governors, expanded the agreement beyond the three main Bay states, and survived some of the twenty-first century's grandest historical dramas of war, recession, and plague. No, the Bay has not been saved, but neither has it been lost. Meanwhile, the restoration movement is in a better position to achieve its goals than ever before. A final review of the Chesapeake's history suggests that over the next fifty years and beyond, challenges will continue to emerge that will test the strength, resolve, and capacity of the Bay's environmental movement. In short, the Bay will always need saving. However, this is not a pessimistic conclusion. Perhaps a future historian will be able to look back and identify the year the Bay was saved. Perhaps that year is closer than we think. More likely, a future historian will instead look back and describe how a broad coalition of people worked together to pass the test of the Chesapeake from one generation to the next.

NOTES

INTRODUCTION

1. For Chesapeake spending since 2014, see https://www.chesapeakeprogress.com/funding; for spending prior to 2014, see Alex Jackson, "Following the money spent on Chesapeake Bay an elusive pursuit," *Capital Gazette* (Annapolis), October 1, 2013, accessed online. For the Everglades, see Anna E. Normand and Pervaze A. Sheikh, "Recent Developments in Everglades Restoration," August 30, 2022, Congressional Research Service, IF11336. For the Great Lakes, see Great Lakes Restoration Initiative website.
2. Arthur Sherwood, "Director's Report," *1977 Annual Report* (Annapolis: CBF, 1978), 3, Chesapeake Bay Foundation internal records, Philip Merrill Environmental Center, Annapolis (hereafter cited as CBF internal records).
3. Thanks to one of my anonymous reviewers for suggesting the connection to Richard White's 1995 book *The Organic Machine: The Remaking of the Columbia River*. Although the field of environmental history has expanded in recent decades to include brilliant scholarship beyond the United States, the field emerged out of a concern with landscapes and issues related to the American West. To give one telling example, six of the first ten books to win the prize for best book in environmental history from the American Society for Environmental History analyzed issues west of the Mississippi. See the essays in "State of the Field: American Environmental History," *Journal of American History* 100, no. 1 (June 2013), for an updated review of important developments in the discipline.
4. "The Charter of Maryland: 1632," accessed online via the Avalon Project, Yale Law School Lillian Goldman Law Library; Wennersten, *Oyster Wars of Chesapeake Bay*.

5. M. J. Langland, *Sediment transport and capacity change in three reservoirs, Lower Susquehanna River Basin, Pennsylvania and Maryland, 1900–2012*, U.S. Geological Survey Open-File Report 2014-1235 (2015).
6. "Vision," *Chesapeake Bay Watershed Agreement*, 2014, accessed online via Chesapeake Bay Program website.
7. U.S. Environmental Protection Agency, *Chesapeake Bay Total Maximum Daily Load for Nitrogen, Phosphorus, and Sediment*, December 29, 2010, accessed online.
8. Felix Morley, quoted in "Informal Minutes of Sixth Organizational Meeting," June 2, 1966, series 1, box 1, folder 1, Chesapeake Bay Foundation records, 0073-MDHC, Special Collections and University Archives, University of Maryland (hereafter cited as CBF archives, UMD).
9. Arthur Sherwood, "Director's Report," *1976 Annual Report* (Annapolis: CBF, 1977), 1, CBF internal records.
10. Ronald Reagan, "Address Before a Joint Session of the Congress on the State of the Union," January 25, 1984, transcript accessed online via the Ronald Reagan Presidential Library and Museum website.
11. "Ban the Phosphates, Help the Bay," editorial, *Washington Post*, February 26, 1985, accessed online.
12. Eugene Cronin, "The Test of the Estuaries," *Bioscience* 20, no. 7 (1970): 395, accession no. 2008-113, box 3, folder 91, L. Eugene Cronin Papers, Special Collections, University of Maryland Libraries (hereafter cited as Cronin Papers).
13. William Cronon, "Reading the Palimpsest," in Curtain, Brush, and Fisher, *Discovering the Chesapeake*, quotation on 372–73.
14. "Bay health somewhat improved with mixed results on trends," *Chesapeake Bay & Watershed 2021 Report Card*, June 6, 2022, University of Maryland Center for Environmental Science, accessed online; *State of the Bay 2022* (Annapolis: CBF, 2023), 12, https://www.cbf.org/document-library/cbf-reports/2022-state-of-the-bay-report.pdf.

1. "SAVE THE BAY"

1. See the USGS website for this and other Chesapeake Bay data.
2. Gottlieb, *Forcing the Spring*, 2. For examples of scholars analyzing national environmentalism from a variety of points of view, see Turner, *Promise of Wilderness*; Hays, *Beauty, Health, and Permanence*; Maher, *Nature's New Deal*; Sellers, *Crabgrass Crucible*; and Rome, *Genius of Earth Day*. Examples of scholarship that specifically analyzes local environmental movements (as opposed to scholarship more generally focused on the environmental history of a specific

locality or region) are fewer, but they include Longhurst, *Citizen Environmentalists*; Huffman, *Protectors of the Land and Water*; Hurley, *Environmental Inequalities*; and Newfont, *Blue Ridge Commons*.

3. Letter to prospective trustees, n.d., series 1, box 1, folder 14, CBF archives, UMD. For instance, Sellers's *Crabgrass Crucible*, Sutter's *Driven Wild*, and Rome's *Bulldozer in the Countryside* show a suburban, terrestrial genesis of environmental concern. Famous works of nature writing that helped raise awareness of ecological issues, such as Aldo Leopold's *Sand County Almanac* or Edward Abbey's *Desert Solitaire*, also take a land-based view. Few writers, scholarly or otherwise, have looked at the emergence of an environmental conscience from an aquatic perspective.

4. All material in this paragraph is from C. A. Porter Hopkins, interview by author, August 21, 2012, Cambridge, MD.

5. Minutes of Fifth Organizational Meeting, May 25, 1966, series 1, box 1, folder 1, CBF archives, UMD; Suzanne Sherwood, interview by author, August 10, 2010, Baltimore.

6. Hopkins interview; Hopkins and Sherwood quotes from "Informal Minutes of Sixth Organizational Meeting," June 2, 1966, series 1, box 1, folder 1, CBF archives, UMD. For CBF's "birthday," see Arthur Sherwood to Trustees, November 29, 1966, series 1, box 1, folder 7, CBF archives, UMD.

7. Marshall Duer to CBF Board of Trustees, December 9, 1966, and Richard Randall to Duer, June 24, 1966, series 1, box 1, folder 7, CBF archives, UMD; Arthur Sherwood to Rolfe Pottberg, February 3, 1967, series 1, box 1, folder 8, CBF archives, UMD.

8. "Treasurer's Report," November 29, 1967, series 1, box 1, folder 2, CBF archives, UMD; G. Howard Gillelan, "Ombudsman for the Bay," *Baltimore Magazine*, October 1968, p. 4, series 3, box 1, folder 2, CBF archives, UMD; Arthur Sherwood, "Director's Report," *1974 Annual Report* (Annapolis: CBF, 1975), 2, CBF internal records.

9. Russell Scott, interview by author, October 26, 2012, Richmond, VA; Arthur Sherwood, "President's Report to Board of Trustees," February 26, 1969, series 1, box 1, folder 3, CBF archives, UMD; Jess Malcolm to Board of Trustees, February 26, 1969, series 1, box 1, folder 2, CBF archives, UMD.

10. Suzanne Sherwood interview; Arthur Sherwood, "Tobe Strong," *CBF News* 6, no. 1 (April 1981): 4, CBF internal records; Arthur Sherwood to CBF Board, December 7, 1967, series 1, box 1, folder 8, CBF archives, UMD.

11. "Minutes—10th Meeting," February 1, 1968; "Minutes—11th Meeting," May 2, 1968; and "Financial Report," April 30, 1968, all in series 1, box 1, folder 2, CBF archives, UMD.

12. "Chesapeake Report," August 15, 1968, CBF internal records.
13. Donald Pritchard to Arthur Sherwood et al., September 6, 1968, series 1, box 5, folder 14, Ferdinand Hamburger Archives of The Johns Hopkins University, Record Group 08.050, Chesapeake Bay Institute archives (hereafter cited as CBI archives, JHU). For more on Pritchard and his contributions to estuarine science, see Kent Mountford, "Donald William Pritchard, known for his research on estuaries, pycnocline, dies," *Bay Journal*, June 1, 1999, accessed online.
14. Pritchard to Sherwood et al., September 6, 1968; Jess Malcolm, "Status of the Bay Report," February 26, 1969, pp. 1–2, series 1, box 1, folder 2, CBF archives, UMD; Hopkins interview.
15. Kirk Davis and Julia King, "Bayonne on the Potomac, Part I: St. Mary's County vs. Steuart Petroleum," *Slackwater* 6 (Spring 2009); Edmund H. Harvey to Jess Malcolm, November 6, 1968, CBF internal records; "Minutes—13th Meeting," November 13, 1968, series 1, box 1, folder 2, CBF archives, UMD; Felix Johnson to Malcolm, October 31, 1968, CBF internal records; Malcolm, "Status of the Bay Report"; Sherwood, "President's Report to Board of Trustees," February 26, 1969.
16. "Financial Report," *1979 Annual Report* (Annapolis: CBF, 1980), 26, CBF internal records; Arthur Sherwood, "CBF Checklist," November 20, 1968, series 1, box 1, folder 9, CBF archives, UMD; Sherwood, "President's Report to Board of Trustees," February 26, 1969, p. 2; Hopkins interview.
17. Gottlieb, *Forcing the Spring*, 2.
18. Portions of the following material are from Ramey, "Calvert Cliffs Campaign, 1967–1971," reprinted with permission of the publisher.
19. Jess Malcolm, statement before the Maryland House of Delegates, Committee on Environmental Matters, March 13, 1969, series 2, box 2, folder 4, CBF archives, UMD; Sherwood to Joseph Tydings, May 10, 1968, series 2, box 1, folder 7, CBF archives, UMD; "Minutes—11th Meeting."
20. For an overview of thermal pollution concerns, see J. Samuel Walker, "Nuclear Power and the Environment"; on L. Eugene Cronin and Joseph A. Mihursky, see "Comments on Proposal by Baltimore Gas and Electric Company for Calvert Cliffs Nuclear Power Plant," February 25, 1970, University of Maryland Natural Resources Institute, series 2, box 2, folder 2, CBF archives, UMD.
21. Untitled press release, Chesapeake Environmental Protection Association, February 10, 1969, series 2, box 2, folder 2, CBF archives, UMD; William Jabine, "The battle to save a bit of green and a bit of blue," *Evening Capital* (Annapolis), April 10, 1969, series 3, box 1, folder 4, CBF archives, UMD.

22. U. S. Atomic Energy Commission, docket nos. 50-317, 50-318, July 8, 1969, pp. 2, 9, series 2, box 2, folder 8, CBF archives, UMD; J. Samuel Walker, *Containing the Atom*, 377.
23. Malcolm to Trustees, June 18, 1969, series 1, box 1, folder 4, CBF archives, UMD; Malcolm to Adolph Ackerman, August 11, 1969, series 2, box 1, folder 9, CBF archives, UMD; Malcolm to Trustees, October 30, 1969, November 21, 1969, and January 13, 1970, all series 1, box 1, folder 4, CBF archives, UMD.
24. Jess Malcolm, statement before the Senate Subcommittee on Intergovernmental Relations, Annapolis, March 3, 1970, p. 3, series 2, box 3, folder 1, CBF archives, UMD; Anthony Z. Roisman, *Petition and Supporting Memorandum*, Calvert Cliffs Coordinating Committee, Inc., June 29,1970, p. 3, series 2, box 2, folder 7, CBF archives, UMD; Malcolm to A.W. Eipper, March 24, 1970, series 2, box 1, folder 11, CBF archives, UMD; National Wildlife Federation, untitled report, June 29, 1970, pp. 3-4, series 2, box 2, folder 7, CBF archives, UMD. On Cayuga Lake, see Balogh, *Chain Reaction*, 262–64; and Nelkin, *Nuclear Power and Its Critics*.
25. Roisman, *Petition and Supporting Memorandum*, 1.
26. Jess Malcolm, "Quad C Newsletter," August 25, 1970, series 2, box 2, folder 2, CBF archives, UMD; Associated Press, "Law suit possible over nuclear plant," *Evening Capital* (Annapolis), November 23, 1970, series 3, box 1, folder 8, CBF archives, UMD; Anthony Z. Roisman, *Petition for Review*, Calvert Cliffs Coordinating Committee, Inc., November 25, 1970, p. 3, series 2, box 2, folder 7, CBF archives, UMD; *Calvert Cliffs Coordinating Committee v. AEC*, 449 F2d 1109 (D.C. Cir. 1971), accessed online.
27. J. Samuel Walker, *Containing the Atom*, 371; Dawson, *Nuclear Power*, 199; Skelly Wright, quoted in Duffy, *Nuclear Politics in America*, 91.
28. Lindstrom and Smith, *National Environmental Policy Act*, 117; Skelly Wright, quoted in Duffy, *Nuclear Politics in America*, 89–90.
29. Duer to CBF Board of Trustees, December 12, 1966, series 1, box 1, folder 7, CBF archives, UMD; Tom Wason, "Malcolm attacks Seaborg," *Evening Capital* (Annapolis), October 31, 1969, series 1, box 1, folder 6, CBF archives, UMD; Sherwood, "CBF Checklist," November 20, 1968, series 1, box 1, folder 9, CBF archives, UMD; Minor Lee Marston, interview by author, September 27, 2012, Annapolis.
30. Jess Malcolm, "The Lemming Psychosis: A special publication for Earth Day 1970," April 22, 1970, series 3, box 1, folder 7, CBF archives, UMD; for the term *great awakening*, Malcolm, quoted in Wayne Hardin, "Malcolm in Midst of Bay Fight," *Evening Sun* (Baltimore), May 22, 1969, series 3, box 1, folder 4, CBF

archives, UMD; Malcolm, "Executive Director's Report—1970," January 27, 1971, series 1, box 1, folder 2, CBF archives, UMD; Rome, *Genius of Earth Day*.

31. Sherwood to Pritchard, December 12, 1970, series 1, box 5, folder 14, CBI archives, JHU; Malcolm, "Executive Director's Report—1970."

32. Malcolm to Norman L. Trott, September 10, 1970, series 1, box 1, folder 11, CBF archives, UMD; Sherwood, "President's Report to Board of Trustees," February 26, 1969; Malcolm, quoted in Wayne Hardin, "Bay Foundation Head Resignation Announced," *Evening Sun* (Baltimore), January 11, 1971, series 3, box 2, folder 1, CBF archives, UMD.

33. Sherwood to Pritchard, December 12, 1970, series 1, box 5, folder 14, CBI archives, JHU; Sherwood, "President's Report to Board of Trustees," February 26, 1969.

34. Arthur Sherwood, "Why We're Doing What We're Doing," *1976 Annual Report* (Annapolis: CBF, 1977), 1, CBF internal records.

2. THE STORM

1. "The Chesapeake Bay Agreement of 1983," December 9, 1983, Records of the Chesapeake Bay Program, Annapolis, available online at https://www.epa.gov/chesapeake-bay-tmdl/chesapeake-bay-agreements.

2. "Bay Program History," Chesapeake Bay Program, accessed February 23, 2020, https://www.chesapeakebay.net/who/bay-program-history.

3. Horton, *Turning the Tide*, xx; Ernst, *Chesapeake Bay Blues*, 14; Governor Hughes, quoted in Ernst, *Chesapeake Bay Blues*, 9.

4. "The History of Chesapeake Bay Cleanup Efforts," Chesapeake Bay Foundation, accessed February 23, 2020, https://www.cbf.org/how-we-save-the-bay/chesapeake-clean-water-blueprint/the-history-of-bay-cleanup-efforts.html.

5. Daniel Pendick, "The Rise and Fall of the Matapeake Monster," *Chesapeake Quarterly* 17, no. 1 (May 2017): 3.

6. Keiner, *Oyster Question*, 1; Kennedy and Breisch, "Sixteen Decades."

7. "Remarks of Hon. George L. Radcliffe," in *Chesapeake Bay Authority*, conference proceedings, October 6, 1933, Baltimore, 9.

8. Wennersten, *Oyster Wars of Chesapeake Bay*; Moore, "Gunfire on the Chesapeake," 367.

9. Rogers C. B. Morton, "Statement for the Governor's Conference on the Chesapeake Bay," in *Proceedings of the Governor's Conference on the Chesapeake Bay* (Queenstown, MD: Wye Institute, 1968), II-176, II-177.

10. Quote from "45th Anniversary of Hurricane Agnes," National Weather Service website; Don Lipman, "Hurricane Agnes: A look back after 40 years,"

Capital Weather Gang (blog), *Washington Post*, June 6, 2012, https://www.washingtonpost.com/blogs/capital-weather-gang/post/hurricane-agnes-a-look-back-after-40-years/2012/06/21/gJQAnDS0sV_blog.html.

11. Simns told his life story to Robert L. Rich Jr., who wrote and published it as *The Best of Times on the Chesapeake Bay: An Account of a Rock Hall Waterman*.
12. Simns, in Rich, *Best of Times*, 180–85.
13. Quotations from this paragraph come from a personal video recording by the journalist Tom Horton. Hargis, Cronin, and Pritchard had gathered for a panel discussion at Tilghman Island on February 17, 1988, with Horton as moderator (hereafter cited as Hargis, Cronin, Pritchard panel discussion). Recognizing the significance of the recollections of these men, Horton arranged to have the discussion recorded. Special thanks to Tom Horton for sharing a copy of the recording with the author.
14. L. Eugene Cronin, "Maryland's Needs for Biological Research Related to the Chesapeake Bay," Natural Resources Institute of the University of Maryland no. 70-114, November 16, 1970, accession no. 2008-113, box 3, folder 86, Cronin Papers.
15. Lynch, "Unprecedented Scientific Community Response to an Unprecedented Event," 31; Hargis, Cronin, Pritchard panel discussion.
16. For more detail, see Lynch, "Unprecedented Scientific Community Response to an Unprecedented Event."
17. C. L. Smith, W. G. MacIntyre, C. A. Lake, and J. G. Windsor Jr., "Effects of Tropical Storm Agnes on Nutrient Flux and Distribution in Lower Chesapeake Bay," in Chesapeake Research Consortium, *Effects of Tropical Storm Agnes on the Chesapeake Estuarine System*, 306.
18. Hargis, Cronin, Pritchard panel discussion.
19. Patti S. Borda, "Sen. 'Mac' Mathias, statesman, leaves legacy in Frederick," *Frederick News-Post*, January 27, 2010, accessed online; Adam Clymer, "Charles Mathias, Former U.S. Senator, Dies at 87," *New York Times*, January 25, 2010, accessed online.
20. Charles Mathias, quoted in "Mathias Begins Bay Fact-Finding Tour with Harbor," *Baltimore Sun*, June 24, 1973, accessed online; Simns, in Rich, *Best of Times*, 198. For Simns's recollection of the banquet, see Rich, *Best of Times*, 241–44.
21. Michael K. Burns, "Mathias Urges U.S., State Work to Save Bay," *Baltimore Sun*, July 10, 1974, accessed online; Tom Horton, "Political, Bureaucratic Lines Tangle Students of the Bay," *Baltimore Sun*, May 23, 1976, accessed online.
22. Chesapeake Research Consortium, *Proceedings of the Bi-State Conference*, vii; Alvin R. Morris, "The Purpose and Nature of the EPA Chesapeake Bay

Program," in Chesapeake Research Consortium, *Proceedings of the Bi-State Conference*, 34.
23. James E. Gutman, "Response to Task Force Report on Non-Point Source Pollution," in Chesapeake Research Consortium, *Proceedings of the Bi-State Conference*, 248.
24. J. Kevin Sullivan et al., "Non-Point Pollution," in Chesapeake Research Consortium, *Proceedings of the Bi-State Conference*, 223, 226.
25. Robert Moore, "Response to Task Force Report on Non-Point Source Pollution," in Chesapeake Research Consortium, *Proceedings of the Bi-State Conference*, 244.
26. "Bi-State Conference on the Bay," editorial, *CBF News* 2, no. 2 (June 1977), CBF internal records.
27. *Report of the Chesapeake Bay Legislative Advisory Commission*, S. Doc. No. 32 (Richmond: Commonwealth of Virginia, 1980), 40.
28. Charles Mathias, "The Challenge of the Chesapeake Bay," reprinted in *Congressional Record*, 94th Cong., 2nd sess., 122, no. 120 (August 5, 1976), 4, accession no. 2008-77, box 8, folder "What Has Mathias Done for the Bay?," Cronin Papers.
29. *Report of the Chesapeake Bay Legislative Advisory Commission*, 41–45.
30. Ernst, *Fight for the Bay*, 10–13. See also Ernst, *Chesapeake Bay Blues*.
31. Rich, *Best of Times*, 200.

3. THE CHESAPEAKE BAY AGREEMENT

1. "The Chesapeake Bay Agreement of 1983," December 9, 1983, Records of the Chesapeake Bay Program, Annapolis, available online at https://www.epa.gov/chesapeake-bay-tmdl/chesapeake-bay-agreements.
2. Ernst, *Fight for the Bay*, xv.
3. Donald D. Deborde to CBF, December 2, 1980, reprinted in *CBF News* 6, no. 1 (April 1981): 7, CBF internal records.
4. Ronald Reagan, "Address Before a Joint Session of the Congress on the State of the Union," January 25, 1984, transcript accessed online via the Ronald Reagan Presidential Library and Museum website.
5. Ronald Reagan, Inaugural Address, January 20, 1981, accessed via reaganlibrary.gov.
6. C. Brandt Short, "Conservation Reconsidered: Environmental Politics, Rhetoric, and the Reagan Revolution," in Peterson, *Green Talk in the White House*, 135.
7. Like the analysis presented here, most scholarly treatments of the Reagan administration and the environment tend to be chapters or sections in larger

works on conservativism and/or environmental policy. For more recent studies, see Nelson, *Nature's Burdens*; and Layzer, *Open for Business*. See also Daynes and Sussman, *White House Politics and the Environment*; Stine, "Natural Resources and Environmental Policy"; Portney, *Natural Resources and the Environment*; Shanley, *Presidential Influence and Environmental Policy*; and V. Smith, *Environmental Policy under Reagan's Executive Order*, for Reagan's influence on environmental policy. For Reagan's influence on the EPA, see Landy, Roberts, and Thomas, *Environmental Protection Agency*; and Mintz, *Enforcement at the EPA*. For a less scholarly account of the first Reagan administration's approach to environmental issues, see Lash, Gilman, and Sheridan, *Season of Spoils*.

8. Skocpol and Williamson, *Tea Party and the Remaking of Republican Conservatism*; M. Smith, *Right Talk*; Crenson and Ginsberg, *Downsizing Democracy*; Julian E. Zelizer, "Seizing Power: Conservatives and Congress since the 1970s," in Pierson and Skocpol, *Transformation of American Politics*; Shaiko, *Voices and Echoes for the Environment*.
9. In 1983 she married Robert F. Buford and took his name, but for consistency and clarity she will be referred to as Anne Gorsuch since that was her name at the time of most of the events discussed here.
10. Layzer, *Open for Business*, 118.
11. Layzer, *Open for Business*, 118; Portney, *Natural Resources and the Environment*, 11.
12. Shanley, *Presidential Influence and Environmental Policy*, 157; Lash, Gilman, and Sheridan, *Season of Spoils*, 7; Landy, Roberts, and Thomas, *Environmental Protection Agency*, 250; Russell Train, "The Destruction of the EPA," editorial, *Washington Post*, February 2, 1982; Andrews, *Managing the Environment, Managing Ourselves*, 259. For data on the budget cuts, see Robert Crandall and Paul Portney, "Environmental Policy," in Portney, *Natural Resources and the Environment*, 66–68.
13. "EPA Program Axed," *CBF News* 6, no. 2 (June 1981), CBF internal records; Shanley, *Presidential Influence and Environmental Policy*, 31–32.
14. Don Baugh, interview by author, August 9, 2012, Annapolis.
15. Arthur Sherwood, "Why We're Doing What We're Doing," *1976 Annual Report* (Annapolis: CBF, 1977), 3, CBF internal records.
16. "Professionals and Radicals," *CBF News* 6, no. 2 (June 1981): 2, CBF internal records; William Baker, "Director's Report," *1981 Annual Report* (Annapolis: CBF, 1982), 4, CBF internal records.
17. Godfrey Rockefeller, "President's Report," and William Baker, "Director's Report," *1982 Annual Report* (Annapolis: CBF, 1983), 1, 4, 6, CBF internal records.

18. "Lehman Replaces White," *CBF News* 4, no. 2 (Summer 1979): 4, CBF internal records; Stuart Lehman, "New Attack on Pollution Problems," *1979 Annual Report* (Annapolis: CBF, 1980), 23, CBF internal records; Lehman, quoted in Claudia Jones, "CBF to Track Major Dischargers," *CBF News* 7, no. 3 (October 1982): 4, CBF internal records.
19. Layzer, *Open for Business*, 132; Landy, Roberts, and Thomas, *Environmental Protection Agency*, 245; William Baker, interview by author, December 6, 2012, Annapolis.
20. Baker interview.
21. John Page Williams, "Appreciation, Understanding are Program Goals: New Concerns and Challenges," *1982 Annual Report* (Annapolis: CBF, 1983), 20, CBF internal records.
22. Anne Gorsuch, quoted in Philip Shabecoff, "House Charges Head of E.P.A. with Contempt," *New York Times*, December 17, 1982, A1. See also D. V. Feliciano, "Gorsuch Cited for Contempt of Congress," *Journal (Water Pollution Control Federation)* 55, no. 2 (1983): 119–22; and Mintz, *Enforcement at the EPA*, 56.
23. *EPA Chesapeake Bay Program: Hearing before the Subcommittee on Governmental Efficiency and the District of Columbia, United States Senate*, 98th Cong. (1983). All subsequent direct quotations from the hearing testimony are from this source.
24. David Hoffman and Cass Peterson, "Buford Quits as EPA Administrator," *Washington Post*, March 10, 1983, A1; Associated Press, "Four More Officials Resign from E.P.A," *New York Times*, May 21, 1983; "Conservationist of the Year," *1984 Annual Report* (Annapolis: CBF, 1985), 7, CBF internal records.
25. Frances Flanigan, "1983 Chesapeake Bay Conference—Status Report," June 1983, series 2, box 3, folder 21, CBI archives, JHU.
26. "Cronkite Leads Bay Talks," *CBF News* 8, no. 3 (December 1983), CBF internal records; Ann Swanson, interview by author, November 13, 2012, Annapolis; Frederick Cusick, "3 Governors Meet to Discuss Cleanup of Chesapeake Bay," *Philadelphia Inquirer*, June 21, 1983, B5.
27. Frances Flanigan, *Choices for the Chesapeake: An Action Agenda—1983 Chesapeake Bay Conference Report*, Citizens Program for the Chesapeake Bay, January 1984; "Cleanup or Whitewash?," editorial, *Washington Post*, December 9, 1983; "Did anything happen?," editorial, *Daily Press* (Newport News, VA), December 14, 1983; Swanson interview.
28. "Cleanup or Whitewash?"; "Did anything happen?"
29. William Ruckelshaus, quoted in Flanigan, *Choices for the Chesapeake*.
30. Marjorie A. Hutter, "The Chesapeake Bay: Saving a National Resource through Multi-State Cooperation," *Virginia Journal of Natural Resources Law* 4, no.2

(1985): 185–207; "States' Bay Initiatives Advance," *CBF News* 9, no. 1 (May 1984): 4, CBF internal records; William Baker, quoted in "Did anything happen?"; Baker interview.

31. Republican aide and Senator Mathias quoted in Alison Muscatine and Sandra Sugawara, "Chesapeake Bay Emerges as Reagan's Environmental Cause," *Washington Post*, October 7, 1984, accessed online; Reagan, State of the Union address, January 25, 1984.

32. "President Ronald Reagan visit the Blackwater National Wildlife Refuge during a Trip to Tour Tilghman Island, Maryland," video footage from Records of the White House Television Office, master tape 73, July 10, 1984, Ronald Reagan Presidential Library website, https://catalog.archives.gov/id/75502438. For an overview of Reagan's trip, see Norman D. Sandler, "President Reagan, starting a three day campaign to improve his . . . ," July 10, 1984, UPI, https://www.upi.com/Archives/1984/07/10/President-Reagan-starting-a-three-day-campaign-to-improve-his/2785458280000/; for criticism, see Steven R. Weisman, "Reagan's Visit to Tidelands Irks Environmental Critics," *New York Times*, July 11, 1984, A16; and Simns in Rich, *Best of Times*, 235.

33. Ronald Reagan to William Baker, July 6, 1984, accession no. 2008-77, box 12, folder "Chesapeake Bay Foundation," Cronin Papers.

34. "Year of the Bay," *CBF News* 8, no. 3 (December 1983): 2, CBF internal records.

35. "Chesapeake Bay Agreement of 1983."

4. PROGRESS AND BACKLASH

1. *1984 Annual Report* (Annapolis: CBF, 1985), 4, 30, CBF internal records; *1994 Annual Report* (Annapolis: CBF, 1995), 30, CBF internal records.

2. U.S. Environmental Protection Agency, *Chesapeake Bay: A Framework for Action*, xv, 8, accessed online; Robert Magnien, Daniel Boward, and Steven Bieber, eds., *The State of the Chesapeake Bay, 1995* (Annapolis: Chesapeake Bay Program, 1995), 1, accessed online.

3. Simns, quoted in Rich, *Best of Times*, 237; bald eagle quote from Russell, *Striper Wars*, 3; William Goldsborough, interview by author, December 3, 2012, Annapolis; striped bass facts from Chesapeake Bay Program website.

4. Juvenile index data, publicly available on the Maryland DNR website, which also has an excellent video explaining this survey; Horton, *Turning the Tide*, 153–60; Goldsborough interview.

5. "Experts Discuss Bay Fisheries," *CBF Newsletter*, May 1975, 1, CBF internal records; Goldsborough interview.

6. "Smart, Smarter, Smartest," editorial, *Baltimore Sun*, July 20, 1981, reprinted in *CBF News* 6, no. 3 (September 1981): 3, CBF internal records; Larry Simns, letter to the editor, *CBF News* 6, no. 4 (November 1981): 2–3, CBF internal records.
7. Russell, *Striper Wars*, 190.
8. Goldsborough interview.
9. "Chesapeake Bay Foundation Testimony Regarding 1985 Maryland Striped Bass Regulations," August 21, 1984, 3, accession no. 2008-77, box 12, folder "Chesapeake Bay Foundation," Cronin Papers; William Goldsborough, "Striped Bass Fishing Ban Announced," *CBF News* 9, no. 3 (November 1984): 1, CBF internal records.
10. "Moratoria Necessary for Rockfish in Our Future," *CBF News* 9, no. 3 (November 1984): 2, CBF internal records; Goldsborough interview; Simns in Rich, *Best of Times*, 240.
11. Goldsborough, "Striped Bass Fishing Ban Announced," 6; "Juvenile Striped Bass Survey," Virginia Institute of Marine Science website.
12. Russell, *Striper Wars*, 165–67.
13. See Maryland DNR website for juvenile index data.
14. Goldsborough interview.
15. Taylor, *Making Salmon*, 238–41.
16. Goldsborough interview.
17. Claiborne W. Gooch III to William Baker, October 24, 1990, accession no. 2008-77, box 10, folder "Alphabetical Files—G," Cronin Papers; Taylor, *Making Salmon*, 241; William Goldsborough to CBF Board of Directors, November 29, 1990, accession no. 2008-77, box 10, folder "Alphabetical Files—G," Cronin Papers.
18. Chesapeake Bay Program, *Chesapeake Bay Striped Bass Fishery Management Plan: Annual Progress Report 1994* (Washington, DC: EPA, February 1995), 1.
19. Chesapeake Bay Program, *The State of the Chesapeake Bay: Third Biennial Monitoring Report—1989* (Annapolis: CBP, 1990), 22.
20. For harvest statistics, see Kennedy, *Shifting Baselines in the Chesapeake Bay*, 43–47; Keiner, *Oyster Question*, 2, 232; and William Goldsborough, "Record Low Harvest Signals Crisis; Future Uncertain for Bay Oysters," *CBF News* 12, no. 1 (May 1987): 8, CBF internal records. See also Livie, *Chesapeake Oysters*.
21. William Baker, "President's Message," *CBF News* 14, no. 4 (December 1989): 2, CBF internal records; Goldsborough interview.
22. William Goldsborough, "Oyster Stocks at Record Low," *CBF News* 16, no. 2 (June 1991): 5, CBF internal records; "Md. Official Rejects Proposal for Moratorium on Oystering," *Washington Post*, June 16, 1991, C11; William Goldsborough,

"Point of No Return; Only drastic action will save the oysters of Chesapeake Bay," editorial, *Washington Post,* July 28, 1991, C8.

23. Larry Simns, "A Moratorium Won't Bring Back the Oyster," editorial, *Washington Post,* August 11, 1991, C8; Goldsborough interview.
24. For profiles on Nick Carter, see Tom Horton, "30 years dedicated to making difference," *Baltimore Sun,* October 13, 2000, accessed online; and Tim Junkin, "Druid of the Chesapeake," *Chesapeake Bay Magazine,* August 16, 2019, accessed online. On Ed Farley, see Amy Pelsinsky, "Living Through History," *Chesapeake Bay Magazine,* November 1, 2018, accessed online; and Hargis, Cronin, Pritchard panel discussion.
25. Hargis, Cronin, Pritchard panel discussion.
26. Horton's description of Carter and the quote from Baugh are from Horton, "30 years dedicated to making a difference"; the other quotes are from Hargis, Cronin, Pritchard panel discussion.
27. William Pruitt, quoted in D'Vera Cohn, "Oyster-Planting Proposal Wins Backing of Va. Marine Institute," *Washington Post,* January 29, 1992, D1; "Oyster Plan Falls Short on Conservation," *CBF News* 18, no. 3 (November 1993): 1, CBF internal records; Goldsborough interview.
28. "Why Sewage Stinks!," *Save the Bay* 28, no. 3 (Summer 2002): 4, CBF internal records; David W. Litke, *Review of Phosphorus Control Measures in the United States and Their Effects on Water Quality,* U.S. Geological Survey, Water-Resources Investigations Report 99-4007, 1991, p. 5, accessed online. Other accounts of this important campaign include Edgar R. Jones and Susan Hubbard, "Maryland's Phosphate Ban—History and Early Results," *Journal (Water Pollution Control Federation)* 58, no. 8 (August 1986): 816–22; Chris Knud-Hansen, *Historical Perspective of the Phosphate Detergent Conflict,* Conflict Research Consortium, University of Colorado, Working Paper 94-54, February 1994, accessed online; and Ernst, *Fight for the Bay,* 83–94.
29. Tom Kenworthy, "Phosphate Ban Advances; Md. Panel Sends Heavily Amended Measure to House," *Washington Post,* March 30, 1985, B1.
30. Baker, quoted in Effie Cottman, "Soap Group Scored," *Capital* (Annapolis), February 18, 1985, 10.
31. Kenworthy, "Phosphate Ban Advances; Md. Panel Sends Heavily Amended Measure to House," *Washington Post,* March 30, 1985, B1; William Baker, quoted in Effie Cottman, "Soap Group Scored," *Capital* (Annapolis), February 18, 1985, 10; Larry Young, quoted in Kenworthy, "Phosphate Ban Advances." For lobbying figures, see R. H. Melton, "Md. Lobbyists May Earn $6 Million; Meal Tabs Total $180,000," *Washington Post,* June 6, 1985, C1; for

CBF's spending, see "Financial Pages," *1985 Annual Report* (Annapolis: CBF, 1986), 31, CBF internal records.

32. Angus Phillips, "Phosphate Fight Begins; Md. Bill Debated Early, at News Conference," *Washington Post*, February 14, 1985, B5.
33. Tom Horton, "Phosphate Issue is not Loss of Jobs," *Baltimore Sun*, March 10, 1985, 1D.
34. See Ernst, *Fight for the Bay*, 90, for this anecdote from Larry LaMotte.
35. Ann Swanson, "'BayWatchers' Are Vital Component of Grassroots Program," *1984 Annual Report*, 14.
36. Ann Swanson, "Your Right to Write," n.d., CBF ephemera in personal collection of Ann Swanson.
37. For more on astroturfing, see Longhurst, *Citizen Environmentalists*, 81–83; E. Walker, *Grassroots for Hire*; Thomas P. Lyon, "Astroturf: Interest Group Lobbying and Corporate Strategy," *Journal of Economics and Management Strategy* 13, no. 4 (Winter 2004): 561; Angus Phillips, "Signature Drive for Money Irks Environmentalists," *Washington Post*, March 7, 1985, B7; Ernst, *Fight for the Bay*, 90–91; and "'CLEAN' is not clean operation," editorial, *Capital* (Annapolis), March 12, 1985, from clippings in personal collection of Ann Swanson.
38. Molly Broderson to petition signers, March 22, 1985, in personal collection of Ann Swanson.
39. Gerald Winegrad, "Badmouthing and falsehoods part of the environmental fight," commentary, *Capital Gazette*, June 10, 2022, published online; Ira C. Cooke, quoted in Kenworthy, "Phosphate Ban Advances."
40. "Phosphate Ban Phone Calls to Delegates," n.d., CBF ephemera in personal collection of Ann Swanson.
41. Jones and Hubbard, "Maryland's Phosphate Ban," 819; R. B. Sellars Jr., D. S. Bauer, J. L. Rein, and M. Jiang, "Effect of Phosphate Detergent Ban on Municipal Treatment Plants in Maryland," (Annapolis: Water Management Administration, Maryland Office of Environmental Program, Division of Municipal Compliance, June 1987); Peter R. Lucchesi, "Virginia Legislature Passes Phosphate Ban Bill," *William & Mary Environmental Law and Policy Review* 12, no. 1 (1987): 15; "Virginia Bans Phosphates," *CBF News* 12, no. 1 (May 1987): 5, CBF internal records.
42. Oreskes and Conway, *Merchants of Doubt*, 9.
43. William Baker, "We *Can* Save the Bay," *1985 Annual Report* (Annapolis: CBF, 1986), 6.
44. Ann Swanson to BayWatchers, October 14, 1986, CBF ephemera in personal collection of Ann Swanson; Ann Powers, "Pennsylvania Office Opens;

Continued Threats to Bay Resources Mark CBF's 20th Year," *1986 Annual Report* (Annapolis: CBF, 1987), 15; "CBF Establishes New Program in Upper Bay," *CBF News* 10, no. 2 (June 1985): 4, CBF internal records; "Richard King Mellon Foundation Supports Pa. Office," *CBF News* 11, no. 1 (May 1986,): 4.

45. "Sexton to Head CBF Pennsylvania Office," *CBF News* 11, no. 3 (October 1986): 8, CBF internal records; Jan Chaplick, "Creek gets high grades from boating students," *Patriot-News* (Harrisburg), May 19, 1987, 10. This was not the first time Pennsylvania students participated in a CBF trip, but it was the first time they did so in Pennsylvania. In 1979 a group of Gettysburg High School students had visited Smith Island, in the middle of the Bay on the Maryland-Virginia border. Prior to 1987, trips involving Pennsylvania students were sporadic. For membership figures and other activities in Pennsylvania, see "The Water and the Land," *1989 Annual Report* (Annapolis: CBF, 1990), 5, CBF internal records.

46. "Gardner To Broaden Ag. Policy Analysis," *CBF News* 11, no. 4 (December 1986,): 8, CBF internal records; Patrick Gardner, "Agriculture and the Bay," *1987 Annual Report* (Annapolis: CBF, 1988), 16, CBF internal records.

47. "CBF Legislative Agendas Summarized," *CBF News* 12, no. 3 (October 1987): 7; "Farmland Preservation Favored by 2–1 Ratio of Voters Statewide," *Philadelphia Inquirer*, November 4, 1987, A14. See also Timothy J. Houseal, "Forever a Farm: The Agricultural Conservation Easement in Pennsylvania," *Dickenson Law Review* 94 (Winter 1990): 527–52; "CRP May Be 'Most Important' Water Quality Program for Agriculture," *CBF News* 13, no. 2 (May 1988): 7; and "Nutrient Pollution is Target of Pennsylvania Plan," *CBF News* 14, no. 2 (June 1989): 12, CBF internal records.

48. William Baker, "President's Report," *1988 Annual Report* (Annapolis: CBF, 1989), 6, CBF internal records; "Nutrient Pollution is Target of Pennsylvania Plan."

49. David Brubaker, quoted in "Conservationist of the Year," *1989 Annual Report*, 32; Ann Swanson, interview by author, November 13, 2012, Annapolis.

50. Karl Blankenship, "Panel: Reducing Nutrients Growing More Difficult," *Bay Journal*, March 1, 1991, accessed online.

51. Tom Horton, "Chesapeake is a victim of Pennsylvania fiasco," *Baltimore Sun*, December 5, 1992, B1.

52. Lamont Garber, quoted in "Interviews: The Sustainable Agriculture Staff," *1992 Annual Report* (Annapolis: CBF, 1993), 7, CBF internal records; Robert Hoyt, quoted in "Nutrient Management Act Signed into Law," *CBF News* 18, no. 2 (July 1993): 8, CBF internal records; "Landmark Coalitions Advance Bay

Protection," *CBF News* 18, no. 2 (July 1993): 7, CBF internal records; Pamela S. Clarke and Stacey M. Cronk, "The Pennsylvania Nutrient Management Act: Pennsylvania Helps to Save the Bay through Nonpoint Source Pollution Management," *Villanova Environmental Law Journal* 6, no. 2 (1995): 319–43.

53. Robert P. Casey, "Pennsylvania Moves to Aid the Bay," *Baltimore Sun*, June 7, 1993, 7A; Hilary Harp Falk, interview by Michael Buckley, *Voices of the Chesapeake Bay*, podcast audio, July 17, 2022; William Baker, "Time to catch the pollution reduction train," *Save the Bay* 30, no. 2 (Spring 2004): 2, CBF internal records; "Promises to Keep," *CBF News* 13, no. 3 (August 1988): 2, CBF internal records; "'Summit Meeting': The Stage is Set," *CBF News* 8, no. 1 (August 1983): 2, CBF internal records.

54. "Chesapeake Bay Agreement: 1992 Amendments," *Chesapeake Bay Program*, August 12, 1992, accessed online.

55. Victoria Churchville, "The Poisoning of Chesapeake Bay," *Washington Post*, June 1, 1986, A1; "The Chesapeake Challenge—Learning not to Pollute," *CBF News* 11, no. 3 (October 1986): 2, CBF internal records; William A. Cook, *Chesapeake Bay Cleanup Program: Hearing before the Subcommittee on Governmental Efficiency and the District of Columbia, United States Senate*, 99th Cong. (1986), 16.

56. "A Pledge of Life for the Bay," editorial, *Washington Post*, December 18, 1987, A26.

57. William Baker, "CBF Urges Stricter Controls to Reduce Pollution," *CBF News* 17, no. 2 (September 1992): 1, CBF internal records.

5. MAKING A NATIONAL TREASURE

1. Jess Malcolm, quoted in "Oil Refinery in St Mary's County Threatens Bay With It's [sic] Pollution," *Kent Island Bay Times*, August 15, 1968; "1987 Chesapeake Bay Agreement," December 15, 1987, Records of the Chesapeake Bay Program, Annapolis, available online at https://www.epa.gov/chesapeake-bay-tmdl/chesapeake-bay-agreements; *Staying the Course*, Chesapeake Bay Commission 2003 Annual Report (Annapolis: CBC, 2004), 27. For the most up-to-date financial data, see https://www.chesapeakeprogress.com/funding.

2. Exec. Order No. 13508, "Chesapeake Bay Protection and Restoration," 74 Fed. Reg. 23099 (May 12, 2009), sec. 201.

3. It should be noted that not all watermen felt this way. David Laird quit the Tangier Sound Watermen's Association in protest over the signs and publicly supported CBF, of which he was a longtime member. See Bill Gifford, "Fire and Water," *Washington Post Magazine*, July 28, 1996, W17.

4. Don Baugh interview by author, August 9, 2012, Annapolis; William Goldsborough, "A Learning Year on Smith Island," *1979 Annual Report* (Annapolis: CBF, 1980), 12, CBF internal records; Gifford, "Fire and Water."
5. Layzer, *Open for Business*, 187–88.
6. Nolan McCarty, "Policy Effects of Polarization," in Pierson and Skocpol, *Transformation of American Politics*; League of Conservation Voters National Environmental Scorecard, February 1996, 3, accessed online.
7. Julian E. Zelizer, "Seizing Power: Conservatives and Congress since the 1970s," in Pierson and Skocpol, *Transformation of American Politics*, 125; Andrews, *Managing the Environment, Managing Ourselves*, 390; "After the Elections: Critical Bay Issues at Stake," *BayWatcher Bulletin*, insert to *Save the Bay*, 20, no. 1 (February 1995): 7, 9, CBF internal records.
8. CBF quote from sidebar in *Save the Bay* 20, no. 2 (July 1995): 2, CBF internal records; "Bud Shuster's Dirty Water Act," editorial, *New York Times*, April 2, 1995, accessed online; "Clean Water Rewrite Stalls in Senate," *CQ Almanac* 1995, 51st ed., 5-5-5-9 (Washington, DC: Congressional Quarterly, 1996), http://library.cqpress.com/cqalmanac/cqal95-1100379.
9. "Congressman Gilchrest on the 'Dirty Water Bill,'" *Save the Bay* 20, no. 2 (July 1995): 1, 3.
10. Susan Carter Brown, "MD General Assembly Update," *Save the Bay* 20, no. 3 (January 1996): 5, CBF internal records; "State Legislative Forecast," *Save the Bay* 22, no. 1 (January 1997): 3, CBF internal records; Atlantic States Marine Fisheries Commission, "ASMFC Presents William Goldsborough Prestigious Captain David H. Hart Award," press release, October 28, 2016; J. Lowell Stoltzfus et al. to Parris N. Glendening, February 6, 1996, accession no. 2008-77, box 10, folder "Alphabetical Files—G," Cronin Papers.
11. William Baker, "Raise Your Voice for the Blue Crab," *Save the Bay* 20, no. 3 (September 1995): 2, CBF internal records; William Goldsborough, interview by author, December 3, 2012, Annapolis.
12. Eugene Cronin to Glendening, February 15, 1996, and William Goldsborough to Cronin, February 21, 1996, both accession no. 2008-77, box 10, folder "Alphabetical Files—G," Cronin Papers.
13. "LCV Announces George Allen First Member of the 2012 'Dirty Dozen,'" press release, January 31, 2012; Bill McAllister, "Interior Official Resigns Under Fire; Assistant Secretary's Personnel Moves Brought Protests From Hill," *Washington Post*, May 27, 1989, A7; D'Vera Cohn, "Allen Appointee Has Detractors; Natural Resources Choice Worries Environmentalists," *Washington Post*, January 24, 1994, D1.

14. Joint Legislative Audit and Review Commission, *Review of the Department of Environmental Quality*, House Doc. 44 (Richmond: Commonwealth of Virginia, 1995), iv; Rex Springston, "Activist Hired to Take State Job; Bourne Hired for Conservation Agency Post Allen Cut in 1994," *Richmond Times-Dispatch*, May 14, 1996, B1; Ellen Nakashima and Spencher S. Hsu, "Official Rips Gilmore Over Memo; GOP Infighting Colors Gubernatorial Contest," *Washington Post*, August 25, 1997, accessed online.
15. D'Vera Cohn, "EPA Says It Will Take Over Va. Air Pollution Program," *Washington Post*, December 1, 1994, D4; Peter Baker, "Allen Sues U.S. on Air Pollution; Governor Contends States Can Decide How to Regulate Range of Areas," *Washington Post*, January 10, 1995, B5; "CBF Action Assures Virginians' Rights to Challenge Permits," *Save the Bay* 21, no. 2 (July 1996): 5; Alan Cooper, "Supreme Court Rejects Virginia Challenge to U.S. Clean Air Law," *Richmond Times-Dispatch*, January 22, 1997, B5; Rex Springston, "Pollution Permit Bill is OK'D," *Richmond Times-Dispatch*, March 1, 1996, A8.
16. Ron Nixon, "Toxic Data Kept Secret; Few Had Access to DEQ Database," *Roanoke (VA) Times*, April 25, 1999, B1; "Hiding Pollution Data Betrays Public Trust," editorial, *Roanoke (VA) Times*, April 28, 1999, A8.
17. Jan Jarrett, "Pennsylvania General Assembly Update," *Save the Bay* 21, no. 1 (January 1996): 6, CBF internal records; "Citizen Advisory Council Appointment," *Save the Bay* 21, no. 3 (October 1996): 5, CBF internal records; "PA General Assembly Update," *Save the Bay* 20, no. 3 (September 1995): 6, CBF internal records; Karl Blankenship, "Two new agencies replace Pennsylvania DER," *Bay Journal*, July 1, 1995, online edition; William Baker, "CBF Fights to Save Wetlands," *Save the Bay* 20, no. 4 (April 1996): 2, CBF internal records.
18. Thomas Grasso, quoted in "Main legislative aim in Maryland assembly: Pfiesteria controls," *Save the Bay* 23, no. 1 (Winter 1998): 3, CBF internal records. On *Pfiesteria*, see Barker, *And the Waters Turned to Blood*; Powlik, *Sea Change*; and Schätzing, *Swarm*.
19. Michael W. Fincham, "How Did a Media Storm Get Started? The Frenzy over *Pfiesteria*," *Chesapeake Quarterly* 6, no. 1 (August 2007): 3. On the ten-year anniversary of the *Pfiesteria* crisis, the entire August 2007 edition of *Chesapeake Quarterly* was devoted to it. See also Fincham, *Pfiesteria Files* (Maryland Public Television, 2007). For a twenty-year retrospective, see Michael Dresser, "Outbreak of toxic *Pfiesteria* in 1997 hurt many—but boosted Chesapeake cleanup," *Baltimore Sun*, July 22, 2017, accessed online. For the $127 million statistic, see Ted Shelsby, "State advertising campaign to tell consumers seafood from Chesapeake Bay is safe; Federal grant aids plan aiming to guard industry from Pfiesteria fear," *Baltimore Sun*, June 10, 1998, 2C.

20. "Pfiesteria generates campaign for farm runoff control," *Save the Bay* 22, no. 2 (Fall 1997): 1, CBF internal records; William Baker, "Pfiesteria's clarion call to action," *Save the Bay* 22, no. 2 (Fall 1997), CBF internal records.
21. *Report of the Citizens Pfiesteria Action Commission*, November 3, 1997, accessed online via Maryland State Archives website, https://msa.maryland.gov/megafile/msa/speccol/sc5300/sc5339/000113/000000/000152/unrestricted/20040010e.html.
22. Boesch, *Cambridge Consensus*; Michael Dresser, "Farm runoff cap in place," *Baltimore Sun*, May 12, 1998, A1; Tom Grasso, quoted in "Fighting Pfiesteria in Maryland," *Save the Bay* 23, no. 2 (Spring 1998): 4, CBF internal records; Robert E. Magnien, "The Dynamics of Science, Perception, and Policy during the Outbreak of *Pfiesteria* in the Chesapeake Bay," *BioScience* 51, no. 10 (October 2001): 851.
23. Lawrence Latane III, "State to Promote Seafood Industry After Microbe Scare," *Richmond Times-Dispatch*, October 23, 1997, B1; John C. Whitehead, Timothy C. Haab, and George R. Parsons, "Economic Effects of Pfiesteria," *Ocean and Coastal Management* 46, no. 9–10 (2003): 845–58; Todd Shields, "Scare Meets Skepticism in VA," *Washington Post*, September 18, 1997, accessed online.
24. "Virginia seeks better permits and public review for feedlots, poultry houses," *Save the Bay* 23, no. 1 (Winter 1998): 5, CBF internal records; "New reforms may reduce nutrients," *Save the Bay* 24, no. 2 (Spring 1999): 4, CBF internal records; "Virginia: Putting nutrients on the agenda," and "CBF Members Make a Difference," *Save the Bay* 23, no. 2 (Spring 1998): 4, 5, CBF internal records.
25. "Pennsylvania works to regulate nutrients," *Save the Bay* 23, no. 2 (Spring 1998): 4, CBF internal records; "New reforms may reduce nutrients," *Save the Bay* 24, no. 2 (Spring 1999): 4, CBF internal records; "Pennsylvania Growing Greener," *Save the Bay* 25, no. 2 (Spring 2000): 4, CBF internal records.
26. James M. Falk, Forbes L. Darby, and Willett Kempton, *Understanding Mid-Atlantic Residents' Concerns, Attitudes, and Perceptions about Harmful Algal Blooms*: Pfiesteria Piscicida, DEL-SG-05-00 (Newark, DE: University of Delaware Sea Grant College Program, August 2000), 40.
27. Francis X. Clines, "Governments Tighten Goals for Restoring Chesapeake," *New York Times*, June 29, 2000, accessed online.
28. Ann Swanson, interview by author, November 13, 2012, Annapolis.
29. Swanson interview; "Chesapeake 2000: Long Range Plan," internal report, accession no. 2008-77, box 12, folder "Loose Files C (1 of 2)," Cronin Papers.
30. "Indicators of the Bay's Health," *1995 Annual Report* (Annapolis: CBF, 1996), 3–7, CBF internal records; "New tool measures progress in Bay restoration effort," *Save the Bay* 23, no. 4 (Fall 1998): 1, CBF internal records; *The State of*

the Bay (Annapolis: CBF, 1998), 2, https://www.cbf.org/document-library/cbf-reports/1998-state-of-the-bay-report.pdf. An important predecessor to the *State of the Bay* report was Tom Horton and William Eichbaum's book *Turning the Tide*, which contained a "report card" on the Bay.

31. "About the Enclosed Annual Report," *Save the Bay* 22, no. 1 (January 1997): 2, CBF internal records; William Baker, "A $50,000 bet for better water," *Save the Bay* 23, no. 1 (Winter 1998): 2, CBF internal records.

32. Chesapeake 2000, June 28, 2000, 8, 11, accessed via Chesapeake Bay Program website; Lawrence Latane III, "Va. Objects to Land Use Goal in Bay Plan; Chesapeake 2000 Draft Expected Today," *Richmond Times-Dispatch*, December 8, 1999, B4; Latane, "Enlarged Bay Cleanup Plan is Celebrated," *Richmond Times-Dispatch*, June 29, 2000, B1.

33. Tom Horton, "Craft of compromise Document," *Baltimore Sun*, June 16, 2000, 2C; William Baker, "New Bay Agreement a step in the right direction," *Save the Bay* 26, no. 3 (Summer 2000): 2.

34. The cost and time frame of the Everglades restoration, like those of the Chesapeake's, have subsequently been revised upward. In August 2022 the Congressional Research Service estimated that CERP will cost $23.2 billion and take until 2050 to implement. See Anna E. Normand and Pervaze A. Sheikh, "Recent Developments in Everglades Restoration," August 30, 2022, Congressional Research Service, IF11336; Clarke and Dalrymple, "$7.8 Billion for Everglades Restoration"; Gonzalez, "Comprehensive Everglades Restoration Plan"; and William Baker, "President's Message" and "What is a Saved Bay worth?," *Save the Bay* 27, no. 2 (Spring 2001): 2, 3, CBF internal records.

35. *The Cost of a Clean Bay: Assessing Funding Needs Throughout the Watershed* (Annapolis: Chesapeake Bay Commission, January 2003); *Seeking Solutions: Chesapeake Bay Commission Annual Report 2001* (Annapolis: CBC, 2002), 34; "Motivating Citizens Throughout the Region," *2001 Annual Report* (Annapolis: CBF, 2002), 15, CBF internal records; "Sewage upgrade bill needs your support," *Save the Bay* 27, no. 3 (Summer 2001): 12, CBF internal records.

36. Gottlieb, *Forcing the Spring*, 407; John Page Williams, interview by author, November 15, 2012, Annapolis; William Baker, "President's Message," *2003 Annual Report* (Annapolis: CBF, January 2004), 1, CBF internal records; Baker, "Challenging a Culture of Failure," *Save the Bay* 33, no. 1 (Spring 2007): 2, CBF internal records.

37. *For the Resources: Chesapeake Bay Commission 2002 Annual Report* (Annapolis: CBC, 2003), 3–4; *Cost of a Clean Bay*, 8. Not until 2007 would the Chesapeake Executive Council publicly acknowledge that the Chesapeake 2000 water

quality goals would not be met by the 2010 deadline. A similar dynamic played out with the 2025 TMDL deadline.

38. "Chesapeake 2000: Long Range Plan"; Williams interview.
39. Williams interview; William Baker, "To save the Bay, business as usual is not acceptable," *Save the Bay* 30, no. 3 (Summer 2004): 2, CBF internal records; Jon Mueller, interview by author, August 7, 2012, Annapolis.
40. "CBF opens federal affairs office" and William Baker, "A line in the Sand," *Save the Bay* 32, no. 1 (March 2006): 1, 8; "Making Environmental Education a National Priority," *Save the Bay* 33, no. 2 (Summer 2007): 20; Baker, "President's Message," *2006 Annual Report* (Annapolis: CBF, 2007), 1; Baker, "Challenging a Culture of Failure," 2; "New Momentum," *2007 Annual Report* (Annapolis: CBF, 2008), 8, all sources from CBF internal records.
41. Steve McAllister, quoted in Isaac Rehert, "Saving the Bay," *Baltimore Sun*, August 12, 1984, People, 6A; all other quotes from Ernst, *Chesapeake Bay Blues*, 137–39.
42. Frances Flanigan, interview by Genevieve de Mahy, recorded in 2019, transcript accessed online via the Peale Museum and IUPUI Ruth Lilly Special Collections and Archives; *Reflections: Chesapeake Bay Commission Annual Report 2004* (Annapolis: CBC, 2005), 16; dates for the formation of Riverkeepers groups obtained from their websites. See also "Find a Bay Organization" on the Chesapeake Bay Program website.
43. "A Thousand New Trees for Monocacy Watershed," *Save the Bay* 33, no. 1 (Spring 2007): 15, CBF internal records; membership and revenue data from CBF's 2001 and 2008 annual reports.
44. William Baker, "Bi-partisanship needed to save the Bay," *Save the Bay* 30, no. 5 (March 2005): 2, CBF internal records.
45. *Reflections: Chesapeake Bay Commission Annual Report 2004*, 36–37; "A Team Effort: Passing Maryland's Historic Sewage Legislation," insert, *Save the Bay* 30, no. 2 (Spring 2004), CBF internal records; Rona Kobell, "Ehrlich to make 'flush tax' law," *Baltimore Sun*, May 23, 2004, accessed online.
46. *Reflections: Chesapeake Bay Commission Annual Report 2004*, 9; "Legislative Victories," *2005 Annual Report* (Annapolis: CBF, 2006), 2–3, CBF internal records; *Focal Points: Chesapeake Bay Commission Annual Report 2006* (Annapolis: CBC, 2007), 11; "New Investment," *2007 Annual Report* (Annapolis: CBF, 2008), 12, CBF internal records.
47. "Demanding Lasting Change," *2006 Annual Report* (Annapolis: CBF, 2007), 2, CBF internal records; "Oyster dredging plan defeated," *Save the Bay* 32, no. 1 (March 2006): 10, CBF internal records; "Record Turn-out for Clean the Bay

Day," *Save the Bay* 33, no. 2 (Summer 2007): 21, CBF internal records. See also *Focal Points: Chesapeake Bay Commission Annual Report 2006;* and *Scaling Up: Chesapeake Bay Commission Annual Report 2007* (Annapolis: CBC, 2008).

48. *Scaling Up: Chesapeake Bay Commission Annual Report 2007*, 25; *2020 REAP Annual Report,* Pennsylvania Department of Agriculture, State Conservation Commission, accessed online.
49. "Innovative agricultural proposal gathers momentum," *Save the Bay* 32, no. 4 (December 2006): 10, CBF internal records; "Victory for REAP," *Save the Bay* 33, no. 3 (Fall 2007): 16, CBF internal records; "A Letter to the President," *Reflections: Chesapeake Bay Commission Annual Report 2004*, 45.
50. Jim DeMint, quoted in Jonathan Weisman, "A Bush Veto Is Overridden for the 1st Time," *Washington Post,* November 9, 2007, accessed online.
51. Ann Powers, "CBF Staff Address Key Problems," *1985 Annual Report* (Annapolis: CBF, 1986), 14; William Baker, quoted in "Help Make the Farm Bill More Bay-Friendly," *Save the Bay* 27, no. 4 (Fall 2001): 12, CBF internal records.
52. *Scaling Up: Chesapeake Bay Commission Annual Report 2007*, 12.
53. "Pollution Reduction," *2008 Annual Report* (Annapolis: CBF, 2009), 2, CBF internal records; "A Fair Share for Bay State Farmers," *Save the Bay* 33, no. 1 (Spring 2007): 12, CBF internal records; "Pollution Reduction," 2; United States Department of Agriculture, *Impacts of Conservation Adoption on Cultivated Acres of Cropland in the Chesapeake Bay Region, 2003–06 to 2011* (Washington, DC: United States Department of Agriculture, December 2013), 6.
54. "Mr. Smith Goes to Washington," *Save the Bay* 34, no. 2 (Summer 2008): 20, CBF internal records.
55. Larry Simns, quoted in "CBF and Allies Begin Legal Action Against the EPA," *Save the Bay* 34, no. 4 (Winter 2008): 25, CBF internal records.
56. "CBF Reports Reveal Need for Toxics Enforcement and Legislation," *Save the Bay* 22, no. 1 (January 1997): 4, CBF internal records; "Agency's failure to protect waters may bring CBF suit," *Save the Bay* 22, no. 3 (Summer 1997): 4, CBF internal records; Chesapeake 2000, 5.
57. Mueller interview.
58. William Baker, "A Message from the President," *2011 Annual Report* (Annapolis: CBF, 2012), 2, CBF internal records. The EPA's website contains a summary of the legal proceedings against the TMDL and links to the court cases discussed above. See "Chesapeake Bay TMDL Court Decisions."
59. "Outreach," and "The Biggest Fight for Clean Water This Nation Has Ever Seen," *2009 Annual Report* (Annapolis: CBF, 2010), 1, 2, CBF internal records; Exec. Order 13508, pt. 1.

60. Arthur Sherwood, "Why We're Doing What We're Doing," *1976 Annual Report* (Annapolis: CBF, 1977), 1, CBF internal records.
61. Mueller interview.

CONCLUSION

1. Gerald Winegrad, "Don't fall for the happy talk: Bay leaders have failed us," *Chesapeake Bay Journal*, January 19, 2023, accessed online.
2. Keiner, *Oyster Question*, 238; Horton, *Turning the Tide*, xvii.
3. Ann Swanson, interview by author, November 13, 2012, Annapolis.
4. Jess Malcolm, "Executive Director's Report—1970," January 27, 1971, series 1, box 1, folder 2, CBF archives, UMD; Arthur Sherwood, "Director's Report," *1978 Annual Report* (Annapolis: CBF, 1979), 2–3; Simns, in Rich, *Best of Times* 194, 196.
5. "Did anything happen?," editorial, *Daily Press* (Newport News, VA), December 14, 1983; *The Cost of a Clean Bay: Assessing Funding Needs Throughout the Watershed* (Annapolis: Chesapeake Bay Commission, January 2003), 8.
6. For Chesapeake spending since 2014, see https://www.chesapeakeprogress.com/funding; and Alex Jackson, "Following the money spent on Chesapeake Bay an elusive pursuit," *Capital Gazette* (Annapolis), October 1, 2013, accessed online.
7. Hilary Harp Falk, interview by Timothy B. Wheeler, "Chesapeake Bay Foundation leader calls for shifts in Bay cleanup," *Chesapeake Bay Journal*, March 6, 2023, accessed online.

BIBLIOGRAPHY

Abbey, Edward. *Desert Solitaire: A Season in the Wilderness.* New York: McGraw-Hill, 1968.

Andersen, Tom. *This Fine Piece of Water: An Environmental History of Long Island Sound.* New Haven, CT: Yale University Press, 2002.

Andrews, Richard N. L. *Managing the Environment, Managing Ourselves: A History of American Environmental Policy.* New Haven, CT: Yale University Press, 2006.

Baliles, Gerald L. *Preserving the Chesapeake Bay.* Martinsville: Virginia Museum of Natural History Foundation, 1995.

Balogh, Brian. *Chain Reaction: Expert Debate and Public Participation in American Commercial Nuclear Power.* Boston: South End, 1979.

Barker, Rodney. *And the Waters Turned to Blood: The Ultimate Biological Threat.* New York: Simon & Schuster, 1997.

Bocking, Stephen. *Nature's Experts: Science, Politics, and the Environment.* New Brunswick, NJ: Rutgers University Press, 2004.

Boesch, Donald. *The Cambridge Consensus: Forum on Land-Based Pollution and Toxic Dinoflagellates in Chesapeake Bay.* Cambridge, MD: University of Maryland Center for Environmental Science, 1997.

Bolster, W. Jeffrey. *The Mortal Sea: Fishing the Atlantic in the Age of Sail.* Cambridge, MA: Belknap Press of Harvard University Press, 2014.

———. "Opportunities in Marine Environmental History." *Environmental History* 11 (2006): 567–97.

Booker, Matthew Morse. *Down by the Bay: San Francisco's History between the Tides.* Berkeley: University of California Press, 2013.

Brady, Lisa M. *War upon the Land: Military Strategy and the Transformation of Southern Landscapes during the American Civil War.* Athens: University of Georgia Press, 2012.

Brait, Susan. *Chesapeake Gold: Man and Oyster on the Bay.* Lexington: University Press of Kentucky, 1990.

Brown, Alexander Crosby. *Steam Packets on the Chesapeake: A History of the Old Bay Line since 1840.* Cambridge, MD: Cornell Maritime, 1961.

Bullard, Robert D. *Dumping in Dixie: Race, Class, and Environmental Quality.* 3rd ed. Boulder, CO: Westview, 2000.

Burke, David G., and Joel E. Dunn, eds. *A Sustainable Chesapeake: Better Models for Conservation.* Arlington, VA: Conservation Fund, 2010.

Carson, Rachel. *Silent Spring.* Boston: Houghton Mifflin, 1962.

Chambers, Erve. *Heritage Matters: Heritage, Culture, History, and Chesapeake Bay.* College Park: University of Maryland Sea Grant Publications, 2006.

Chesapeake Research Consortium. *The Effects of Tropical Storm Agnes on the Chesapeake Estuarine System.* Baltimore: Johns Hopkins University Press, 1976.

———. *Proceedings of the Bi-State Conference on the Chesapeake Bay, April 27–29, 1977.* Richmond: Commonwealth of Virginia, 1977.

Christofferson, Bill. *The Man from Clear Lake: Earth Day Founder Senator Gaylord Nelson.* Madison: University of Wisconsin Press, 2004.

Clark, Ray, and Larry Canter. *Environmental Policy and NEPA: Past, Present, and Future.* Boca Raton, FL: St. Lucie, 1997.

Clarke, Alice L., and George H. Dalrymple. "$7.8 Billion for Everglades Restoration: Why Do Environmentalists Look So Worried?" *Population and Environment* 24, no. 6 (2003): 541–69.

Cohen, Michael P. *The History of the Sierra Club, 1892–1976.* San Francisco: Sierra Club Books, 1988.

Cole, Luke W., and Sheila R. Foster. *From the Ground Up: Environmental Racism and the Rise of the Environmental Justice Movement.* New York: NYU Press, 2000.

Cook, Constance Ewing. *Nuclear Power and Legal Advocacy: The Environmentalists and the Courts.* Lexington, MA: Lexington Books, 1981.

Cowdrey, Albert. *This Land, This South: An Environmental History.* Lexington: University Press of Kentucky, 1983.

Crenson, Matthew A., and Benjamin Ginsberg. *Downsizing Democracy: How America Downsized Its Government and Sidelined Its Citizens.* Baltimore: Johns Hopkins University Press, 2002.

Cronon, William. *Changes in the Land: Indians, Colonists, and the Ecology of New England.* New York: Hill & Wang, 2003.

———. *Nature's Metropolis: Chicago and the Great West.* New York: W.W. Norton, 1992.

———, ed. *Uncommon Ground: Toward Reinventing Nature.* New York: W.W. Norton, 1995.

Curtain, Philip D., Grace S. Brush, and George W. Fisher, eds. *Discovering the Chesapeake: The History of an Ecosystem.* Baltimore: Johns Hopkins University Press, 2001.

Davis, Jack E. *An Everglades Providence: Marjory Stoneman Douglas and the American Environmental Century.* Athens: University of Georgia Press, 2009.

Davison, Steven G., Jay G. Merwin Jr., John Capper, Garrett Power, and Frank R. Shivers Jr. *Chesapeake Waters: Four Centuries of Controversy, Concern, and Legislation.* 2nd ed. Centreville, MD: Tidewater, 1997.

Dawson, Frank G. *Nuclear Power: The Development and Management of a Technology.* Seattle: University of Washington Press, 1976.

Daynes, Byron W., and Glen Sussman. *White House Politics and the Environment: Franklin D. Roosevelt to George W. Bush.* College Station: Texas A&M University Press, 2010.

Dewey, Scott Hamilton. *Don't Breathe the Air: Air Pollution and U.S. Environmental Politics, 1945–1970.* College Station: Texas A&M University Press, 2000.

Dorbin, Ann E., and Richard A. K. Dorbin. *Saving the Bay: People Working for the Future of the Chesapeake.* Baltimore: Johns Hopkins University Press, 2001.

Drake, Brian Allen. *Loving Nature, Fearing the State: Environmentalism and Antigovernment Politics before Reagan.* Seattle: University of Washington Press, 2015.

Duffy, Robert J. *Nuclear Politics in America: A History and Theory of Government Regulation.* Lawrence: University Press of Kansas, 1997.

Dunlap, Thomas R. *DDT: Scientists, Citizens, and Public Policy.* Princeton, NJ: Princeton University Press, 1981.

Ernst, Howard R. *Chesapeake Bay Blues: Science, Politics, and the Struggle to Save the Bay.* Lanham, MD: Rowman & Littlefield, 2003.

———. *Fight for the Bay: Why a Dark Green Environmental Awakening is Needed to Save the Chesapeake Bay.* Lanham, MD: Rowman & Littlefield, 2009.

Fox, Stephen R. *The American Conservation Movement: John Muir and His Legacy.* Boston: Little, Brown, 1981.

Gibbons, Boyd. *Wye Island: Insiders, Outsiders, and Change in a Chesapeake Community.* Special reprint ed. New York: Routledge, 2007.

Gisolfi, Monica R. *The Takeover: Chicken Farming and the Roots of American Agribusiness.* Athens: University of Georgia Press, 2017.

Gonzalez, George A. "The Comprehensive Everglades Restoration Plan: Environmental or Economic Sustainability?" *Polity* 37, no. 4 (2005): 466–90.

Gottlieb, Robert. *Forcing the Spring: The Transformation of the American Environmental Movement*. 2nd ed. Washington, DC: Island, 2005.

Grunwald, Michael. *The Swamp: The Everglades, Florida, and the Politics of Paradise*. New York: Simon & Schuster, 2006.

Hays, Samuel P. *Beauty, Health, and Permanence: Environmental Politics in the United States, 1955–1985*. New York: Cambridge University Press, 1989

———. *Conservation and the Gospel of Efficiency: The Progressive Conservation Movement, 1890–1920*. Cambridge, MA: Harvard University Press, 1959.

Hedeen, Robert A. *The Oyster: The Life and Lore of the Celebrated Bivalve*. Centreville, MD: Tidewater, 1986.

Horton, Tom. *Turning the Tide: Saving the Chesapeake Bay*. 2nd ed. Washington, DC: Island, 2003.

Horton, Tom, and William Eichbaum. *Turning the Tide: Saving the Chesapeake Bay*. Washington, DC: Island, 1991.

Houde, Edward D. *Managing the Chesapeake's Fisheries: A Work in Progress*. College Park: Maryland Sea Grant College Publications, 2011.

Huffman, Thomas R. *Protectors of the Land and Water: Environmentalism in Wisconsin, 1961–1968*. Chapel Hill: University of North Carolina Press, 1994.

Hurley, Andrew. *Environmental Inequalities: Class, Race, and Industrial Pollution in Gary, Indiana, 1945–1980*. Chapel Hill: University of North Carolina Press, 1995.

Jacoby, Karl. *Crimes against Nature: Squatters, Poachers, Thieves, and the Hidden History of American Conservation*. Berkeley: University of California Press, 2001.

Jasper, James M. *Nuclear Politics: Energy and the State in the United States, Sweden, and France*. Princeton, NJ: Princeton University Press, 1990.

Joppke, Christian. *Mobilizing Against Nuclear Energy: A Comparison of Germany and the United States*. Berkeley: University of California Press, 1993.

Keiner, Christine. *The Oyster Question: Scientists, Watermen, and the Maryland Chesapeake Bay since 1880*. Athens: University of Georgia Press, 2009.

Kennedy, Victor S. *Shifting Baselines in the Chesapeake Bay: An Environmental History*. Baltimore: Johns Hopkins University Press, 2018.

Kennedy, Victor S., and Linda L. Breisch. "Sixteen Decades of Political Management of the Oyster Fisher in Maryland's Chesapeake Bay." *Journal of Environmental Management* 16 (1983): 153–71.

Klingle, Matthew W. *Emerald City: An Environmental History of Seattle*. New Haven, CT: Yale University Press, 2007.

Landy, Mark K., Marc J. Roberts, and Stephen R. Thomas. *The Environmental Protection Agency: Asking the Wrong Questions*. New York: Oxford University Press, 1990.

Lash, Jonathan, Katherine Gilman, and David Sheridan. *A Season of Spoils: The Reagan Administration's Attack on the Environment.* New York: Pantheon Books, 1984.

Layzer, Judith A. *Open for Business: Conservatives' Opposition to Environmental Regulation.* Cambridge, MA: MIT Press, 2012.

Leopold, Aldo. *A Sand County Almanac: And Sketches Here and There.* New York: Oxford University Press, 1949.

Lindstrom, Matthew, and Zachary Smith. *The National Environmental Policy Act: Judicial Misconstruction, Legislative Indifference, and Executive Neglect.* College Station: Texas A&M University Press, 2001.

Livie, Kate. *Chesapeake Oysters: The Bay's Foundation and Future.* Charleston, SC: History Press, 2015.

Longhurst, James Lewis. *Citizen Environmentalists.* Medford, MA: Tufts University Press, 2010.

Lynch, M. P. "An Unprecedented Scientific Community Response to an Unprecedented Event: Tropical Storm Agnes and the Chesapeake Bay." In *Hurricane Isabel in Perspective,* ed. K. G. Sellner, 29–36. Edgewater, MD: Chesapeake Research Consortium, 2005.

Maher, Neil M. *Nature's New Deal: The Civilian Conservation Corps and the Roots of the American Environmental Movement.* New York: Oxford University Press, 2008.

Matuszeski, William. *Inquiry in a Culture of Consensus: Science and Management for the Chesapeake Bay.* College Park: University of Maryland Sea Grant Publications, 2008.

McEvoy, Arthur F. *The Fisherman's Problem: Ecology and Law in the California Fisheries, 1850–1980.* New York: Cambridge University Press, 1986.

McNeill, J. R. *Something New Under the Sun: An Environmental History of the Twentieth-Century World.* New York: W.W. Norton, 2000.

Meindl, Christopher F. "Past Perceptions of the Great American Wetland: Florida's Everglades during the Early Twentieth Century." *Environmental History* 5, no. 3 (2000): 378–95.

Merchant, Carolyn. *Ecological Revolutions: Nature, Gender, and Science in New England.* Chapel Hill: University of North Carolina Press, 1989.

Middleton, Arthur Pierce. *Tobacco Coast: A Maritime History of Chesapeake Bay in the Colonial Era.* Baltimore: Johns Hopkins University Press and the Maryland State Archives, 1984.

Milazzo, Paul Charles. *Unlikely Environmentalists: Congress and Clean Water, 1945–1972.* Lawrence: University Press of Kansas, 2006.

Miller, Char. *Gifford Pinchot and the Making of Modern Environmentalism.* Washington, DC: Island, 2001.

Mintz, Joel A. *Enforcement at the EPA: High Stakes and Hard Choices*. Austin: University of Texas Press, 1995.

Moore, James. "Gunfire on the Chesapeake: Governor Cameron and the Oyster Pirates, 1882–1885." *Virginia Magazine of History and Biography* 90, no. 3 (1982): 367–77.

Nash, Roderick. *Wilderness and the American Mind*. 4th ed. New Haven, CT: Yale University Press, 2001.

Nelkin, Dorothy. *Nuclear Power and Its Critics: The Cayuga Lake Controversy*. Ithaca, NY: Cornell University Press, 1971.

Nelson, Daniel. *Nature's Burdens: Conservation and American Politics, the Reagan Era to the Present*. Logan: Utah State University Press, 2017.

Newfont, Kathryn. *Blue Ridge Commons: Environmental Activism and Forest History in Western North Carolina*. Athens: University of Georgia Press, 2012.

Opie, John. *Ogallala: Water for a Dry Land*. Lincoln: University of Nebraska Press, 1993.

Oreskes, Naomi, and Erik M. Conway. *Merchants of Doubt: How a Handful of Scientists Obscured the Truth on Issues from Smoking to Global Warming*. New York: Bloomsbury, 2010.

Paolisso, Michael. *Chesapeake Environmentalism: Rethinking Culture to Strengthen Restoration and Resource Management*. College Park: University of Maryland Sea Grant Publications, 2006.

Peters, Margaret T. *Conserving the Commonwealth: The Early Years of the Environmental Movement in Virginia*. Charlottesville: University of Virginia Press, 2008.

Peterson, Tarla Rai, ed. *Green Talk in the White House: The Rhetorical Presidency Encounters Ecology*. College Station: Texas A&M University Press, 2004.

Petulla, Joseph M. *American Environmentalism: Values, Tactics, Priorities*. College Station: Texas A&M University Press, 1980.

Pierson, Paul, and Theda Skocpol, eds. *The Transformation of American Politics: Activist Government and the Rise of Conservatism*. Princeton, NJ: Princeton University Press, 2007.

Portney, Paul, ed. *Natural Resources and the Environment: The Reagan Approach*. Washington, DC: Urban Institute, 1984.

Powlik, James. *Sea Change*. New York: Island Books, 1999.

Ramey, Andrew S. "The Calvert Cliffs Campaign, 1967–1971: Protecting the Public's Right to Knowledge." In *Nuclear Portraits: Communities, the Environment, and Public Policy*, ed. Laurel MacDowell, 121–48. Toronto: University of Toronto Press, 2017.

Reiger, John F. *American Sportsmen and the Origins of Conservation*. 3rd ed. Corvallis: Oregon State University Press, 2007.

Rich, Robert L., Jr. *The Best of Times on the Chesapeake Bay: An Account of a Rock Hall Waterman*. Atglen, PA: Schiffer, 2012.

Roland, John V., Glenn E. Moore, and Michael A. Bellanca. "The Chesapeake Bay Oil Spill—February 2, 1976: A Case History." *International Oil Spill Conference Proceedings* 1977, no. 1 (March 1977): 523–27.

Rome, Adam. *The Bulldozer in the Countryside: Suburban Sprawl and the Rise of American Environmentalism*. New York: Cambridge University Press, 2001.

———. *The Genius of Earth Day: How a 1970 Teach-In Unexpectedly Made the First Green Generation*. New York: Hill & Wang, 2013.

Rothman, Hal K. *Saving the Planet: The American Response to the Environment in the Twentieth Century*. Chicago: Ivan R. Dee, 2000.

Rozwadowski, Helen M., and David K. van Keuren, eds. *The Machine in Neptune's Garden: Historical Perspectives on Technology and the Marine Environment*. Sagamore Beach, MA: Science History Publications, 2004.

Russell, Dick. *Striper Wars: An American Fish Story*. Washington, DC: Island, 2006.

Sabin, Paul. *The Bet: Paul Ehrlich, Julian Simon, and Our Gamble over Earth's Future*. New Haven, CT: Yale University Press, 2013.

Sanders, Jeffrey C. *Seattle and the Roots of Urban Sustainability: Inventing Ecotopia*. Pittsburgh: University of Pittsburgh Press, 2010.

Schätzing, Frank. *The Swarm*. New York: William Morrow Paperbacks, 2007.

Schrepfer, Susan R. *The Fight to Save the Redwoods: A History of Environmental Reform, 1917–1978*. Madison: University of Wisconsin Press, 1983.

Sellers, Christopher. *Crabgrass Crucible: Suburban Nature and the Rise of Environmentalism in Twentieth Century America*. Chapel Hill: University of North Carolina Press, 2012.

Shabecoff, Philip. *A Fierce Green Fire: The American Environmental Movement*. Rev. ed. Washington, DC: Island, 2003.

Shaiko, Ronald G. *Voices and Echoes for the Environment: Public Interest Representation in the 1990s and Beyond*. New York: Columbia University Press, 1999.

Shanley, Robert A. *Presidential Influence and Environmental Policy*. Westport, CT: Greenwood, 1992.

Sherwood, Arthur. *Understanding the Chesapeake: A Layman's Guide*. Centerville, MD: Tidewater, 1973.

Silver, Timothy. *A New Face on the Countryside: Indians, Colonists, and Slaves in the South Atlantic Forests, 1500–1800*. New York: Cambridge University Press, 1990.

Siry, Joseph V. *Marshes of the Ocean Shore: Development of an Ecological Ethic*. College Station: Texas A&M University Press, 1984.

Skocpol, Theda, and Vanessa Williams. *The Tea Party and the Remaking of Republican Conservatism*. New York: Oxford University Press, 2013.

Smith, Kimberly K. *African American Environmental Thought: Foundations*. Lawrence: University Press of Kansas, 2007.

Smith, Mark A. *The Right Talk: How Conservatives Transformed the Great Society into the Economic Society*. Princeton, NJ: Princeton University Press, 2007.

Smith, V. Kerry, ed. *Environmental Policy under Reagan's Executive Order: The Role of Benefit-Cost Analysis*. Chapel Hill: University of North Carolina Press, 1984.

Steinberg, Theodore. *Nature Incorporated: Industrialization and the Waters of New England*. New York: Cambridge University Press, 2003.

Stewart, Mart. *"What Nature Suffers to Groe": Life, Labor, and Landscape on the Georgia Coast, 1680–1920*. Athens: University of Georgia Press, 1996.

Stine, Jeffrey K. "Natural Resources and Environmental Policy." In *The Reagan Presidency: Pragmatic Conservatism and Its Legacies*, ed. W. Elliot Brownlee and Hugh Davis Graham, 233–56. Lawrence: University Press of Kansas, 2003.

Straughan, Baird, and Tom Pollak. *The Broader Movement: Nonprofit Environmental and Conservation Organizations, 1989–2005*. Washington, DC: Urban Institute, 2008.

Summers, Gregory. *Consuming Nature: Environmentalism in the Fox River Valley, 1850–1950*. Lawrence: University Press of Kansas, 2006.

Sutter, Paul. *Driven Wild: How the Fight against Automobiles Launched the Modern Wilderness Movement*. Seattle: University of Washington Press, 2005.

Sutter, Paul, and Christopher J. Manganiello. *Environmental History and the American South: A Reader*. Athens: University of Georgia Press, 2009.

Sutter, Paul, and Paul M. Pressly, eds. *Coastal Nature, Coastal Culture: Environmental Histories of the Georgia Coast*. Athens: University of Georgia Press, 2018.

Sze, Julie. *Noxious New York: The Racial Politics of Urban Health and Environmental Justice*. Cambridge, MA: MIT Press, 2006.

Tarr, Joel A., ed. *Devastation and Renewal: An Environmental History of Pittsburgh and Its Region*. Pittsburgh: University of Pittsburgh Press, 2005

———. *The Search for the Ultimate Sink: Urban Pollution in Historical Perspective*. Akron, OH: University of Akron Press, 1996.

Taylor, Joseph E., III. *Making Salmon: An Environmental History of the Northwest Fisheries Crisis*. Seattle: University of Washington Press, 2001.

Turner, James Morton. *The Promise of Wilderness: American Environmental Politics since 1964*. Seattle: University of Washington Press, 2012.

Uekötter, Frank, ed. *The Turning Points of Environmental History*. Pittsburgh: University of Pittsburgh Press, 2010.

U.S. Environmental Protection Agency. *Chesapeake Bay: A Framework for Action*. Washington, DC: GPO, 1983.

———. *Chesapeake Bay Program: Findings and Recommendations*. Washington, DC: GPO, 1983.

———. *Chesapeake Bay Program Technical Studies: A Synthesis.* Washington, DC: GPO, 1982.

Vileisis, Ann. *Discovering the Unknown Landscape: A History of America's Wetlands.* Washington, DC: Island, 1997.

Walker, Edward. *Grassroots for Hire: Public Affairs Consultants in American Democracy.* New York: Cambridge University Press, 2014.

Walker, J. Samuel. *Containing the Atom: Nuclear Regulation in a Changing Environment, 1963–1971.* Berkeley: University of California Press, 1992.

———. "Nuclear Power and the Environment: The Atomic Energy Commission and Thermal Pollution, 1965–1971." *Technology and Culture* 30, no. 4 (1989): 964–92.

Warner, William W. *Beautiful Swimmers: Watermen, Crabs and the Chesapeake Bay.* 2nd ed. New York: Back Bay Books, 1994.

Warren, Louis S. *The Hunter's Game: Poachers and Conservationists in Twentieth-Century America.* New Haven, CT: Yale University Press, 1997.

Way, Albert G. *Conserving Southern Longleaf: Herbert Stoddard and the Rise of Ecological Land Management.* Athens: University of Georgia Press, 2011.

Wellock, Thomas R. *Critical Masses: Opposition to Nuclear Power in California, 1958–1978.* Madison: University of Wisconsin Press, 1998.

Wennersten, John R. *Chesapeake: An Environmental Biography.* Baltimore: Maryland Historical Society, 2001.

———. *The Oyster Wars of Chesapeake Bay.* Centreville, MD: Tidewater, 1981.

Weyler, Rex. *Greenpeace: How a Group of Journalists, Ecologists, and Visionaries Changed the World.* New York: Rodale, 2004.

White, Richard. "Historiographical Essay: American Environmental History; The Development of a New Field," *Pacific Historical Review* 54 (1985): 297–335.

———. *The Organic Machine: The Remaking of the Columbia River.* New York: Hill & Wang, 1995.

Worster, Donald. *Dust Bowl: The Southern Plains in the 1930s.* New York: Oxford University Press, 1979.

———, ed. *The Ends of the Earth: Perspectives on Modern Environmental History.* New York: Cambridge University Press, 1988.

———. *Rivers of Empire: Water, Aridity, and the Growth of the American West.* New York: Pantheon, 1985.

Zelko, Frank. *Make It a Green Peace! The Rise of Countercultural Environmentalism.* New York: Oxford University Press, 2013.

INDEX

Italicized page numbers refer to figures and maps.

accountability, 28, 139, 158, 160
AEC. *See* Atomic Energy Commission (AEC)
Agnes. *See* Tropical Storm Agnes (1972)
Agnew, Spiro, 55
agriculture: CBF policy on, 131–36; Farm Bill, 133, 177–79, 182, 184; industrial, 2, 103, 179; land preservation for, 132; pollution from, 7, 60–62, 68, 93, 98, 130–37, 154–57, 175; as stressor on Chesapeake Bay's ecosystem, 3, 103; sustainability of, 134, 136; water quality and, 7, 60, 133
air pollution, 149, 163
Allen, George, 148–50, 156, 165
Alliance for the Chesapeake Bay, 66, 101, 170
American Society for Environmental History, 197n3
Andrews, Richard, 144
antienvironmentalism, 12, 75–79, 86, 139–40, 142–51, 164, 166, 195

astroturfing, 125, 210n37
Atlantic States Marine Fisheries Commission (ASMFC), 109–12, 114, 146–48
Atlantic Striped Bass Conservation Act of 1984, 111, 114
Atomic Energy Commission (AEC), 31, 33, 35–42

Baker, James, 92
Baker, William: on blue crab regulations, 147; on CBF in Pennsylvania, 130, 133, 137, 151; on Chesapeake Bay Agreement (1992), 139, 140; on Chesapeake Bay Watershed Litigation Project, 168; on Chesapeake 2000 agreement, 163–66, 169; on CLEAN's deception, 127; on EPA study of Chesapeake Bay, 86, 88–90; on Farm Bill and agricultural conservation, 178; on nutrient pollution, 134; on oyster fishery,

Baker, William (*continued*): 117, 161; on *Pfiesteria piscicida* crisis, 154; on phosphate bans, 124; on politics of postponement, 172; on public support for Chesapeake environmentalism, 83, 98; Reagan's letter of support to, 100; rise to CBF leadership, 81–82; testimony before Congress, 88–90, 89; on TMDL provisions, 182
bald eagles, 3, 99, 106
Baliles, Gerald, 128, 138
Baltimore Gas & Electric (BGE), 31–32, 35–36, 38–39, 41–42
Barry, Marion, 138
Baugh, Don, 81, 82, 85, 120, 143
BayWatchers, 123–25, 127–29
Bell, Brad, 152–53
BGE. *See* Baltimore Gas & Electric (BGE)
Bibko, Peter N., 88–93, 97
Biden, Joe, 192
Bi-State Conference on Chesapeake Bay (1977), 66–70
Blackwater National Wildlife Refuge, 2–3, 99, 174, 207n32
Bliley, Thomas, Jr., 145
blue crabs, 13, 35, 94, 147–48, 162
Boehlert, Sherwood, 178
Brewster, Daniel, 63
Broderson, Molly, 126–27
Brown, Susan Carter, 146
Brown, Torrey, 110, 111, 114–15
Brubaker, David, 133–34
Buford, Robert F., 205n9
Burkholder, JoAnn, 152–53
Bush, George H. W., 143
Bush, George W.: antienvironmentalism of, 142, 164, 166; backlash against environmental policy approach, 174, 177; budget cuts to Chesapeake Bay restoration, 169; Chesapeake Bay Commission's letter to, 176; failure to issue executive order on Chesapeake Bay, 183, 192; war on terror and, 166

Callahan, Debra, 144
Calvert, Lord (Baron of Baltimore), 6
Calvert Cliffs Coordinating Committee (Quad-C), 33, 37–42
Calvert Cliffs Coordinating Committee, Inc. v. U.S. Atomic Energy Commission (1971), 33, 39–40
Calvert Cliffs nuclear power plant: CBF's campaign against, 31–33, 35–43; CEPA's opposition to, 35, 37, 45; construction permit for, 35–36, 38; environmental impact statement for, 38–40; foreshadowing of future trends in, 10; once-through cooling system proposal, 33, 35; thermal pollution and, 32, 33, 35–36, 200n20
Carson, Rachel, 101
Carter, Nick, 119–21, 209n24
Casey, Robert, Sr., 134–36, 139
CBC. *See* Chesapeake Bay Commission (CBC)
CBF. *See* Chesapeake Bay Foundation (CBF)
CBI. *See* Chesapeake Bay Institute (CBI)
CBL. *See* Chesapeake Biological Laboratory (CBL)
CEPA (Chesapeake Environmental Protection Association), 35, 37, 45

CERP. *See* Comprehensive Everglades Restoration Plan (CERP)
Chafee, John, 86, 111, 145
Charles I (king of England), 6
Chesapeake Bay: dead zones within, 7, 189; economic importance of, 1, 2, 7, 12; environmental significance of, 1–3; EPA study on decline of, 11, 49–50, 52, 65–69, 86–93, 105, 192; historical and cultural importance of, 1–2, 141; integrity of ecosystem, 3, 23, 28, 115, 138, 161; key locations within, 34; report cards on health of, 14, 189, 216n30; sustainability of, 3, 8, 51, 105; USGS data on, 17, 198n1; watershed overview, 4–8, 5
Chesapeake Bay Agreement (1983): criticisms of, 74–75, 102; EPA and, 49, 86, 93; executive council formed by, 95; objectives of, 49, 74, 102; signing ceremony, 93, 96; strengths and weaknesses of, 103; Tropical Storm Agnes and, 53; as turning point in Chesapeake environmentalism, 48, 74–75, 93, 97, 157
Chesapeake Bay Agreement (1987), 138–39, 141, 154, 158–59
Chesapeake Bay Agreement (1992), 139, 140, 159
Chesapeake Bay Commission (CBC): analysis of federal funding, 141, 167, 192; Chesapeake 2000 agreement and, 165, 166; Choices for the Chesapeake Conference and, 93–97; creation of, 50, 69–72, 93; lack of regulatory authority, 71–72; letter to George W. Bush, 176; membership expansion, 7, 73, 94–95; on nutrient pollution reduction, 172–73; Waterkeepers Chesapeake and, 170
Chesapeake Bay Foundation (CBF): agricultural policy, 131–36; backlash against, 142–43; BayWatchers and, 123–25, 127–29; birth of, 10, 19–22, 65, 185; Bi-State Conference on Chesapeake Bay and, 67, 70; Calvert Cliffs campaign, 31–33, 35–43; on Chesapeake Bay Program, 80–81; on Clean Water Amendments, 145; cost estimate for Chesapeake 2000 agreement, 164–65; critiques of, 169–70; environmental defense program, 124; environmental education program, 29, 82, 87, 100, 131–32, 169; funding for, 22–25, 28–30, 42–44, 85, 104, 171; hate mail received by, 75–76, 100; indicator goals formulated by, 159–62; litigation program, 12, 168–69, 172, 177–83; membership of, 25, 28, 43, 66, 83–85, 84, 104, 130–31, 161, 162, 171; on oyster fishery, 117–19, 121; panel discussions organized by, 60; in Pennsylvania, 130–37, 150–51, 211n45; philosophical shift within, 133; phosphate bans and, 123–28, 126; Piney Point campaign and, 25–32, 41, 43; Policy and Scientific Review Committee, 26; pollution-enforcement programs, 84–85; Reagan's letter of support for, 100; regional approach of, 10, 18, 28–30, 40, 81, 130, 170, 193; report card on health of Chesapeake Bay, 14, 160; "Save the Bay" bumper stickers from, 19, *19*, 22, 98; statement of goals in charter

Chesapeake Bay Foundation (*continued*):
of, 83; on striped bass, 107–14;
tension within, 23, 26–27, 31, 33,
41–43; transition period for, 81–86;
Tylerton facility arson (1995),
142–43, 147

Chesapeake Bay Hydraulic Model, 52, 61

Chesapeake Bay Institute (CBI), 26,
55, 59, 61

Chesapeake Bay Program: biennial
monitoring reports, 105; Citizens Advisory Committee, 133;
creation of, 3, 11, 48, 50, 192; EPA
and, 11, 80–81, 90–93, 102, 158; list
of organization connected with
Chesapeake Bay, 171; management
structure of, 93; membership of, 7,
11, 73, 137; Reagan administration
and, 80–81, 83, 99; Republican efforts to undo accomplishments of,
12; on striped bass, 114; voluntary
nature of, 51

Chesapeake Bay Watershed Litigation
Project, 168–69

Chesapeake Biological Laboratory
(CBL), 13, 53, 55, 59, 61

Chesapeake environmentalism: backlash against, 12, 75–77, 104, 105,
139–40, 142–51; CBF's relationship
to, 29–31, 101; citizen activism
and, 51, 59, 72, 105, 123–25; costs
related to, 1, 18, 164, 193, 197n1;
dominant narrative of, 47–53, 49,
69; expansion to Pennsylvania,
129–37; false starts and dead ends,
40–44, 52–56, 60; future outlook
for, 194–96; Horton's metaphor
for, 188–89, 192; number of groups
related to, 21; *Pfiesteria piscicida*
crisis and, 142, 151, 154–59, 161,
181, 193; politics of, 9, 11, 72, 79,
100–101, 143–44, 154; progression from state to national issue,
190–93; public support for, 83, 85,
95–96, 98, 105, 186, 194; Reagan
and, 11, 76–77, 79–83, 85, 98–101;
as regional movement, 44–45, 73,
123, 129, 142, 158, 186; setbacks for,
12, 102, 134, 142; shortcomings of,
3, 9, 50–51, 187–89; Tropical Storm
Agnes and, 11, 48, 53, 66, 73, 192;
turning points in, 48, 74–75, 93, 97,
102, 154, 157, 192; value of historical
analysis of, 3, 189–90

Chesapeake Environmental Protection
Association (CEPA), 35, 37, 45

Chesapeake Healthy and Environmentally Sound Stewardship of Energy
and Agriculture Act of 2007
(CHESSEA), 178–79

Chesapeake Research Consortium
(CRC), 59, 62, 69

Chesapeake 2000 agreement: cost
estimate for, 164–65; drafting process, 159, 161; failure of, 158, 163–67,
176, 216–17n37; implementation
and enforcement, 165–67, 169;
indicator goals adopted by, 160–63;
Pfiesteria piscicida crisis as catalyst
for, 142, 158–59, 181

Chesapeake Watershed Agreement
(2014), 8, 194–95

CHESSEA (Chesapeake Healthy and
Environmentally Sound Stewardship of Energy and Agriculture Act
of 2007), 178–79

Chinchilli, Jolene, 151
Choices for the Chesapeake Conference (1983), 93–97
citizen activism, 51, 59, 72, 105, 123–25
Citizens Program for the Chesapeake Bay. *See* Alliance for the Chesapeake Bay
Clean Air Act Amendments of 1990, 143
CLEAN (Consumers League for Environmental Action Now), 125–27, 129
Clean Water Act of 1972, 79, 145, 168, 169, 177, 180–82
Clean Water Amendments of 1995 ("Dirty Water Bill"), 145–46
climate change, 3, 56, 67, 143, 185, 194
Clinton, Bill, 145
Cole, Kenneth J., 94–95
Collins, Charles H., 89
commercial fishing: for blue crabs, 147; for oysters, 117–19; as stressor on Chesapeake Bay's ecosystem, 3, 103; for striped bass, 106, 108–10, 113, 114
Commonwealth of Virginia v. State of Maryland (2003), 6
Comprehensive Everglades Restoration Plan (CERP), 1, 164–65, 197n1, 216n34
Conservation Reserve Program (CRP), 132–33
Consumers League for Environmental Action Now (CLEAN), 125–27, 129
Contract with America, 143–44, 168
Conway, Erik, 129
Cook, William A., 138
Cousteau, Jacques, 95, 96
Coy, Jeffrey, 134

crabs. *See* blue crabs
CRC (Chesapeake Research Consortium), 59, 62, 69
Cronin, L. Eugene, 13–14, 35, 59–62, 119–21, 147–48, 203n13
Cronkite, Walter, 94
Cronon, William, 13–14
CRP (Conservation Reserve Program), 132–33
Cummings, Bill, 99, 102

dead zones, 7, 189
Deborde, Donald, 75–77
Delaware: Chesapeake Watershed Agreement and, 8; regional conference on Chesapeake Bay (1933), 53–55
DeMint, Jim, 177
DEQ (Virginia Department of Environmental Quality), 148–50
dermo, 115, 116, 118, 122
Dingell, John, 178
"Dirty Water Bill" (Clean Water Amendments of 1995), 145–46
DNR. *See* Maryland Department of Natural Resources (DNR)
Duer, Marshall, 22
Dunlop, Becky Norton, 148–50

Earth Day, 10, 13, 19, 41–44, 48, 66, 101–2
Ehrlich, Robert, 172, 174
Eichbaum, William, 216n30
Endangered Species Act of 1973, 79
environmental education, 29, 82, 87, 100, 131–32, 158, 169

environmental impact statements, 38–40
environmentalisms. *See* Chesapeake environmentalism; U.S. environmentalisms
environmental politics, 9, 11, 72, 79, 100–101, 143–44, 154
environmental reasonableness, 23, 82, 133
EPA. *See* U.S. Environmental Protection Agency (EPA)
Ernst, Howard, 50, 72, 169–70
erosion, 60, 133, 154, 175
eutrophication, 7, 122, 123
Everglades, 1, 2, 164–65, 169, 197n1, 216n34

Falk, Hilary Harp, 136–37, 194–95
Family Farm Movement, 135
Farley, Ed, 119–21, 209n24
Farm Bill, 133, 177–79, 182, 184
farming. *See* agriculture
Fincham, Michael, 152–53
fishing: overfishing, 107, 108, 114, 119, 147; recreational, 106–13, 119, 154, 180. *See also* commercial fishing; specific species
fish kills, 12, 35, 152–57
Flanigan, Frances, 101, 170
Fowler, Bernie, 154–55, 180, 182
Frosh, Brian, 155

Garber, Lamont, 134
Gardner, Patrick H., 131–33, 135
Gardner, Richard R., 89
Gibson Island Club, 20, 21, 24

Gilchrest, Wayne, 145–46, 178
Gilmore, James, 149, 156, 162, 163
Gingrich, Newt, 143–44
Glendening, Parris, 147–48, 154–55
Global Climate Protection Act of 1986, 79
Goldsborough, William, 106, 108–10, 112–14, 117–19, 122, 143, 146–48
Gooch, Claiborne W., III, 113–14
Gore, John, 32
Gorsuch, Anne: backlash against, 78, 80, 94, 101; CBF on, 82, 83; confirmation as EPA administrator, 79; in contempt of Congress, 79, 87–88; on EPA study of Chesapeake Bay, 80, 86–88; marriage of, 205n9; resignation from EPA, 79, 86, 92
Governor's Conference on Chesapeake Bay (1968), 55, 65
Grasso, Tom, 152, 155
Great Lakes Restoration Initiative, 1, 197n1
Great Recession, 176, 180
Gutman, James, 68
Gwaltney of Smithfield, Ltd. v. Chesapeake Bay Foundation, Inc. (1987), 85

Hargis, William, 59–62, 119–21, 203n13
Harvey, Edmund, 28
Hopkins, C. A. Porter, 19, 20–22, 27, 29, 185
Horton, Tom: on Chesapeake 2000 agreement, 163; on Mathias's failed proposal, 66; metaphor for Chesapeake environmentalism, 188–89, 192; on Nutrient Management Act, 135; panel discussion convened by,

119–21, 203n13; on phosphate bans, 124; *Turning the Tide*, 50, 216n30
Hoyt, Robert, 136
Hughes, Harry, 50, 93–94, 96, 97, 101, 155, 180
hunting, 20, 30, 53

Jackson, Lisa, 183
Johns Hopkins University, 26, 55, 59
Johnson, Felix, 28

Kaine, Tim, 174, 183
Keiner, Christine, 53, 115–16, 188
Kennedy, Ted, 111
Killius, Anna, 194
Kimmich, Bill, 131
Kind, Ron, 178

Laird, David, 212n3
laundry detergent phosphate ban, 123–29, *126*, 137
Layzer, Judith, 79, 143
League of Conservation Voters, 144, 148
Lehman, Stuart, 84–85
Liddy, G. Gordon, 149
lobbyists: astroturf groups, 125; Clean Water Amendments and, 145; Nutrient Management Act and, 133, 134; phosphate bans and, 123–28; spending by, 124, 209n31; striped bass issues and, 108

Malcolm, Jess: Calvert Cliffs campaign and, 31–33, 35–38, 41–43; on Chesapeake Bay as national issue, 141, 191; environmental philosophy of, 23, 27, 133; ouster from CBF, 31, 33, 40–44, 82, 172; Piney Point campaign and, 25–29, 31, 41, 43; on proving worth of CBF, 24; Quad-C formed by, 33, 37, 41; retirement from public life, 41, 44
Mansfield, Mike, 63
Maroon, Joseph, 128, 157
Marston, Minor Lee, 41
Maryland: antienvironmentalism in, 146–48; Bi-State Conference on Chesapeake Bay (1977), 66–70; blue crab monitoring program, 147; Chesapeake Bay Agreement and (1983), 49, 86, 102; Chesapeake Bay Commission and, 50, 69–73, 93; Chesapeake 2000 agreement and, 161; Choices for the Chesapeake Conference (1983), 93–97; economic gain from Chesapeake Bay, 7; environmental education in, 87, 132; financial commitment to Chesapeake Bay recovery, 98, 167; Governor's Conference on Chesapeake Bay (1968), 55, 65; oyster fishery in, 6, 53, 55, 115–18, *116*, 121–22; *Pfiesteria piscicida* crisis in, 152–55; phosphate ban in, 123–29, *126*; postwar development in, 20–21; Potomac River conflicts with Virginia, 6; regional conference on Chesapeake Bay (1933), 53–55; striped bass fishery in, 106–14; wastewater treatment plant upgrades in, 172–73
Maryland Charterboat Association, 113

Maryland Department of Natural Resources (DNR), 106–7, 109–10, 113, 115, 119–21, 147–48, 207n4
Maryland Saltwater Sportfishermen's Association, 113, 180
Maryland-Virginia Compact (1785), 6
Maryland Watermen's Association (MWA), 58–59, 64, 101, 106, 108, 113, 118, 180, 191
Mathes, Ruth, 37
Mathias, Charles McCurdy "Mac," Jr.: award received by, 49; as CBF board member, 65, 85–86; in dominant narrative of Chesapeake environmentalism, 48–53; EPA study and, 11, 49–50, 52, 65–66, 86, 88, 90–92, 192; family and educational background, 63; on New Federalism program, 98; Republican Party and, 63, 78, 101; retirement of, 64, 65, 138, 144, 168; Ruckelshaus supported by, 93; Simns's relationship with, 64, 73, 203n20; Title II commission proposal, 70; tour of Chesapeake Bay by, 64–65
McAllister, Steve, 170
McCarty, Nolan, 144
McGrath, David, 81
menhaden, 114, 153, 175
Mikulski, Barbara, 178
Moore, Robert, 69
Morgan, Cranston, 49
Morley, Felix, 10
Morone saxatilis. *See* striped bass (*Morone saxatilis*)
Morris, Alvin, 67, 69
Morton, Rogers C. B., 55–56, 64–65, 185

MSX, 115, 116, 118, 122
Mueller, Jon, 168–69, 177, 180–82, 185–86
Murphy, Tayloe, 180
Murray, E. Churchill, 37
MWA. *See* Maryland Watermen's Association (MWA)

National Audubon Society, 22, 85
National Environmental Policy Act of 1970 (NEPA), 31, 33, 37–40, 45
National Wildlife Federation (NWF), 37–38
Nature Conservancy, 169, 171
New York, Chesapeake Watershed Agreement and, 8
NIMBY (not in my backyard), 29–31, 35, 38, 45
nitrogen pollution: agricultural, 130; eutrophication and, 7, 123; public understanding of, 154; reduction strategies, 179; TMDL for, 8–9, 12
Nixon, Richard, 65, 93
Nixon, Ron, 150
nuclear power plant. *See* Calvert Cliffs nuclear power plant
Nuclear Regulatory Commission, 40
Nutrient Management Act of 1993, 130, 133–36, 139, 157, 175
nutrient pollution: agricultural, 60–62, 68, 130, 132–34, 155–57; EPA research on, 50, 67–69; eutrophication and, 123; land-use changes and, 62; *Pfiesteria piscicida* crisis and, 152, 154–57; reduction goals, 138–39, 154, 159, 161, 172–73; TMDL for, 8–9, 180. *See also* nitrogen pollution; phosphorous pollution

NWF (National Wildlife Federation), 37–38

Obama, Barack, 142, 182–85, 192, 193
oil refinery. *See* Piney Point oil refinery
Oreskes, Naomi, 129
Osborne, Carl, 89
overfishing, 107, 108, 114, 119, 147
oysters, 115–22; Chesapeake Bay Foundation on, 117–19, 121; commercial fishing for, 117–19; diseases affecting, 115–18, 121, 122; harvest statistics, 116, *116*, 118, 189, 208n20; as keystone species, 115; methods for harvesting, 53, 120; overharvesting concerns, 117, 118, 121; Oyster Wars, 6, 55; population collapse, 115–18, 121–22; reefs/beds for, 7, 106, 115, 117, 161, 164, 174–75

Pennsylvania: agricultural pollution from, 7, 93, 130–37, 175; antienvironmentalism in, 150–51; CBF in, 130–37, 150–51, 211n45; Chesapeake Bay Agreement and (1983), 49, 74, 86, 93, 102; Chesapeake Bay Commission and, 7, 73, 94–95; Chesapeake Bay Program and, 7, 73, 137; Chesapeake 2000 agreement and, 161; Choices for the Chesapeake Conference (1983), 93–97; environmental education in, 131; expansion of Chesapeake environmentalism to, 129–37; financial commitment to Chesapeake Bay recovery, 98, 157, 167; *Pfiesteria piscicida* crisis in, 157; phosphate ban in, 128, 129, 137; wastewater treatment plant upgrades in, 173
Pennsylvania Association for Sustainable Agriculture, 136
periodization, usefulness of, 47, 194
Pfiesteria piscicida crisis (1997): Cambridge Consensus report on, 155; Chesapeake environmentalism and, 142, 151, 154–59, 161, 181, 193; economic impact of, 154, 156; fish kills resulting from, 12, 152–57; media coverage of, 152–53, 214n19; menhaden with lesions associated with, 153; nutrient pollution and, 152, 154–57; sensationalist works related to, 152
phosphate bans, 123–29, *126*, 137
phosphorous pollution: agricultural, 130; eutrophication and, 7, 123; prevention efforts, 123–29; public understanding of, 154; TMDL for, 8–9, 12, 180
Piney Point oil refinery, 25–32, 41, 43
politics: of the "base," 144; community organizing and, 21; environmental, 9, 11, 72, 79, 100–101, 143–44, 154; macroevents within, 82; polarization of, 18, 78, 88; of postponement, 172; regional, 63–66; striped bass and, 108; of war on terror, 165. *See also* Republican Party
pollution: agricultural, 7, 60–62, 68, 93, 98, 130–37, 154–57, 175; air, 149, 163; CBF enforcement programs, 84–85; monitoring programs, 150; nitrogen, 7–9, 12, 130, 154, 179, 180; oyster harvests impacted by, 121;

pollution (*continued*):
permits for, 84–85, 149; phosphorous, 7–9, 12, 123–30, 154, 180; prevention efforts, 122–29; sediment, 7–9, 12, 58, 62, 117, 132, 175, 180; small craft, 60–61; thermal, 32, 33, 35–36, 60, 200n20. *See also* nutrient pollution

population growth, 3, 103, 187

Potomac River, 2, 6, 25, 148, 155

Potomac River Association (PRA), 29–30

Powers, Ann, 130

Pritchard, Donald, 26–27, 41, 44, 59–63, 119, 121, 200n13

Pruitt, William, 121

Quad-C (Calvert Cliffs Coordinating Committee), 33, 37–42

Radcliffe, George, 53–55

Rambo, Sylvia, 183

Randall, Richard, 22–23

Reagan, Ronald: antienvironmentalism of, 77–79, 86, 144, 195; Chesapeake environmentalism and, 11, 76–77, 79–83, 85, 98–101; environmental politics and, 11, 79, 100–101; EPA during presidency of, 78–83, 86–93, 205n7; literature review on environmental policy, 204–5n7; New Federalism program introduced by, 98; State of the Union address (1984), 11, 76–77, 99; visit to Chesapeake Bay, 99, 207n32

REAP (Resource Enhancement & Protection) program, 175–76

recreational fishing, 106–13, 119, 154, 180

Republican Party: antienvironmentalism of, 12, 79, 139–40, 142–51; conservatism of, 12, 63, 142–44, 149, 150; Contract with America and, 143–44, 168; moderates within, 63, 78, 85–86, 145

Ridge, Tom, 151, 163

Rivanna Conservation Alliance, 171

Riverkeeper organizations, 170–71, 173, 176, 217n42

Robb, Charles S., 93–94, 96, 98, 101

Rockefeller, Godfrey, 49

rockfish. *See* striped bass

Rome, Adam, 42

Ruckelshaus, William D., 80, 86, 92–93, 97–99, 101

Russell, Dick, 106

Safe Drinking Water Act of 1974, 79

salmon fisheries, 112–13

Santa Barbara oil spill (1969), 27

Sarbanes, Paul, 86, 88, 91–93, 165, 174

Scott, Bobby, 178

Scranton, William, III, 96

Seaborg, Glenn, 41

sediment pollution: agricultural, 132, 175; eutrophication and, 7; land-use changes and, 62; oyster harvests impacted by, 117; TMDL for, 8–9, 12, 180; from Tropical Storm Agnes, 58

September 11, 2001 terrorist attacks, 165, 166

Sexton, Thomas P., III, 131–33, 136

Sherwood, Arthur: Calvert Cliffs campaign and, 32–33, 36, 41, 42; on Chesapeake Bay as overlooked ecosystem, 2, 66; environmental philosophy of, 23, 82, 133; frustration expressed by, 191; on funding for CBF, 23–25, 28–29, 42–44; Mathias's Chesapeake Bay tour and, 65; ouster of Malcolm from CBF, 33, 42–44, 82; on Piney Point campaign, 27; role in CBF's founding, 20–22, 185

Sherwood, Suzanne, 21

Showers, John R., 94–95

Shuster, Bud, 145

Sierra Club, 37, 38, 85, 132, 170

Siglin, Doug, 169, 177–79

Simns, Larry: on agricultural pollution, 68; *The Best of Times on the Chesapeake Bay*, 58; Maryland Watermen's Association and, 58–59, 64, 101, 106, 108, 118, 180, 191; Mathias's relationship with, 64, 73, 203n20; on oyster fishery, 118, 121; on Reagan's visit to Chesapeake Bay, 99; on striped bass, 106, 108–10, 113; on TMDL provisions, 180; on Tropical Storm Agnes, 58–59, 63

Stafford, Robert, 86

Steuart Petroleum Company, 25–28, 30

Stoltzfus, J. Lowell, 147

striped bass (*Morone saxatilis*), 106–15; ASMFC management plan, 109–12; Chesapeake Bay Foundation on, 107–14; commercial fishing for, 106, 108–10, 113, 114; harvesting bans/reductions, 110–15; juvenile index, 107, 107–12, 114, 189, 207n4; legal size limits for, 107–9, 111–13; overfishing concerns, 107, 108, 114; population collapse, 106, 108–10; recreational fishing for, 106, 108–10, 113; sustainability of fishery, 109, 112

Strong, C. Trowbridge "Tobe," 24, 28–30, 43, 44

Strong, L. Corrin, 24, 25, 28

Studds, Gerry, 111

suburbanization, 3, 7, 68, 103, 132

Sullivan, J. Kevin, 68

Superfund program, 79, 80, 87, 88

Susquehanna River, 4, 6–7, 17, 56, 130, 131

sustainability: of agriculture, 134, 136; of blue crab fishery, 162; of Chesapeake Bay, 3, 8, 51, 105; of oyster harvests, 116; of striped bass fishery, 109, 112

Swanson, Ann, 94–96, 124–25, 127–29, 134, 158–59, 190

Tangier Sound Watermen's Association, 143, 212n3

Taylor, Joseph, 112

thermal pollution, 32, 33, 35–36, 60, 200n20

Thomas, Lee, 138

Thornburgh, Dick, 94, 95, 101

Tilghman, Dick, 95

Total Maximum Daily Load (TMDL), 8–9, 12, 180–86, 188, 193, 194, 218n58

Train, Russell, 65, 80

Tropical Storm Agnes (1972): Chesapeake environmentalism and, 11, 48, 53, 66, 73, 192; evolution of Chesapeake science and, 59–63; oyster harvests impacted by, 116–17; rainfall resulting from, 56, 57, 61; sediment pollution following, 58

Trump, Donald, 192

University of Maryland: Center for Environmental Science, 14; Chesapeake Biological Laboratory, 13, 53, 55, 59, 61

U.S. Army Corps of Engineers, 52, 61

U.S. environmentalisms: 1970s as "environmental decade," 12; backlash against, 12, 75–79, 101, 105, 139–40, 142–45; CBF's relationship to, 29–31, 101; genesis of, 20, 199n3; literature review, 198–99n2; radicalism and, 82–83; Reagan and, 77–81, 85, 204–5n7; regional variation of, 18, 45; turning points for, 101–2

U.S. Environmental Protection Agency (EPA): annual water quality studies, 91; Bi-State Conference on Chesapeake Bay and, 67–69; budget for, 79–80, 99, 205n12; Chesapeake Bay Agreement and (1983), 49, 86, 93; Chesapeake Bay Program and, 11, 80–81, 90–93, 102, 158; Chesapeake 2000 agreement and, 161; Chesapeake Watershed Agreement and, 8; Federal Leadership Committee chaired by, 184; lack of early involvement in Chesapeake environmentalism, 10–11; Reagan's impact on, 78–83, 86–93, 205n7; study on decline of Chesapeake Bay, 11, 49–50, 52, 65–69, 86–93, 105, 192; TMDL for pollution established by, 8–9, 12, 182–83

U.S. Geological Survey (USGS), 17, 198n1

U.S. Supreme Court, 6, 9, 85, 137, 150, 183, 193–94

Utermohl, Edward, 32

Van Hollen, Chris, 178

VIMS. *See* Virginia Institute of Marine Science

Virginia: air-pollution permit process in, 149; antienvironmentalism in, 148–50; Bi-State Conference on Chesapeake Bay (1977), 66–70; blue crab monitoring program, 147; Chesapeake Bay Agreement and (1983), 49, 86, 102; Chesapeake Bay Commission and, 50, 69–73, 93; Chesapeake 2000 agreement and, 161; Choices for the Chesapeake Conference (1983), 93–97; economic gain from Chesapeake Bay, 7; environmental education in, 87; financial commitment to Chesapeake Bay recovery, 98, 156, 167; Maryland-Virginia Compact (1785), 6; oyster fishery in, 6, 55, 115, 117–18, 121–22, 174–75; *Pfiesteria piscicida* crisis in, 156–57; phosphate ban in, 128, 129; Potomac River conflicts with Maryland, 6; regional conference on Chesapeake Bay (1933), 53–55; striped bass fishery in, 107, 108, 111, 112; toxic pollution monitoring program, 150; wastewater treatment plant upgrades in, 173–74

Virginia Department of Environmental Quality (DEQ), 148–50

Virginia Institute of Marine Science (VIMS), 53, 55, 59, 111, 119, 121, 147, 150

Walker, J. Samuel, 36

Warner, John, 98–99, 101

Warner, William, 95
war on terror, 164–66, 180
Washington, DC: Chesapeake Bay Agreement and (1983), 49, 102; Chesapeake 2000 agreement and, 161; Chesapeake Watershed Agreement and, 8; phosphate ban in, 128, 129; regional conference on Chesapeake Bay (1933), 53–55
wastewater treatment plants, 128, 156, 172–74
waterfowl, 20, 26, 30, 53, 54
Waterkeepers Chesapeake, 170
watermen. *See* commercial fishing
water quality: agricultural practices and, 7, 60, 133; Chesapeake 2000 goals for, 216–17n37; EPA's annual studies of, 91; oyster fishery and, 116–18, 122; phosphate bans and, 124, 129; report cards for assessment of, 14; Simns on changes related to, 64; stabilization of, 105. *See also* pollution
Water Quality Improvement Act of 1998, 155

Water Resources Development Act of 2007, 177, 179
Water Resources Planning Act of 1965, 70
Watt, James, 78, 101
Weicksel, Allen, 135, 136
Wenger, Noah, 134
West Virginia, Chesapeake Watershed Agreement and, 8
wetlands: destruction of, 21, 31, 60, 151, 154; importance to estuarine ecosystem, 13; restoration efforts, 17, 151, 161, 162, 164; state and federal protection of, 145, 151, 174
Wilderness Society, 85
Williams, Anthony, 180
Williams, John Page, 87
Windjammers, 20–22
Winegrad, Gerald, 127, 188, 190
Wolff, George, 134
Works Progress Administration (WPA), 53
Wright, J. Skelly, 39–41

Young, Larry, 124

www.ingramcontent.com/pod-product-compliance
Lightning Source LLC
Chambersburg PA
CBHW030202250225
22505CB00003B/154